MAKING HABITS, BREAKING *HABITS*

MAKING **HABITS** **BREAKING** *HABITS*

How to Make Changes that Stick

JEREMY DEAN

ONEWORLD

A ONEWORLD BOOK

Published in Great Britain and the Commonwealth by Oneworld Publications 2013

First published in the United States by Da Capo Press,
a member of the Perseus Books Group

A CIP record for this title is available from
the British Library

ISBN 978-1-85168-989-7
eISBN 978-1-78074-217-5

Cover design by Richard Green
Typeset by Littler Ascending
Printed and bound by Nørhaven A/S, Denmark

Oneworld Publications
10 Bloomsbury Street, London WC1B 3SR, England

To Howard and Patricia

For in truth habit is a violent and treacherous schoolmistress. She establishes in us, little by little, stealthily, the foothold of her authority; but having by this mild and humble beginning settled and planted it with the help of time, she soon uncovers to us a furious and tyrannical face against which we no longer have the liberty of even raising our eyes.

Montaigne

CONTENTS

MAKING HABITS, BREAKING *HABITS*

PART I

ANATOMY OF A HABIT

1

BIRTH OF A HABIT

This book started with an apparently simple question that seemed to have a simple answer: How long does it take to form a new habit? Say you want to go to the gym regularly, eat more veg, learn a new language, make new friends, practise a musical instrument, or achieve anything that requires regular application of effort over time. How long should it take before it becomes a part of your routine rather than something you have to force yourself to do?

I looked for an answer the same way most people do nowadays: I asked Google. The search suggested the answer was clear-cut. Most top results made reference to a magic figure of 21 days. These websites maintained that 'research' (and the scare-quotes are fully justified) had found that if you repeated a behaviour every day for 21 days, then you would have established a brand-new habit. There wasn't much discussion of what type of behaviour it was, or the circumstances you had to repeat it in, just this figure of 21 days. Exercise, smoking, writing a diary, or turning cartwheels; you name it, 21 days is the answer. In addition, many authors recommend that it's crucial to maintain a chain of 21 days without breaking it. But where does this number come from? Since I'm a psychologist with research training, I'm used to seeing references that would support a bold statement like this. There were none.

My search turned to the library. There, I discovered a variety of stories going around about the source of the number. Easily my favourite concerns a plastic surgeon, Maxwell Maltz MD. Dr Maltz published a book in 1960 called *Psycho-Cybernetics* in which he noted that amputees took, on average, 21 days to adjust to the loss of a limb, and he argued that people take 21 days to adjust to any major life change.[1] He also wrote that he saw the same pattern in those whose faces he had operated on. He found that it took about 21 days for their self-esteem either to rise to meet their newly created beauty or stay at its old level.

The figure of 21 days has exercised an enormous power over self-help authors ever since. Bookshops are filled with titles like *Millionaire Habits in 21 Days*, *21 Days to a Thrifty Lifestyle*, *21 Days to Eating Better*, and finally, the most optimistic of all: *21-Day Challenge: Change Almost Anything in 21 Days* (at least it acknowledges that it might be a challenge!). Occasionally, the 21-day period is deemed a little too optimistic and we are given an extra week to transform ourselves. These more generous titles include *The 28-Day Vitality Plan* and *Diet Rehab: 28 Days to Finally Stop Craving the Foods that Make You Fat.*

Whether 21 or 28 days, it's clear that what we eat, how we spend money, or indeed anything else we do, has little in common with losing a leg or having plastic surgery. To take Dr Maltz's observations of his patients and generalize them to almost all human behaviour is optimistic at best. It's even more optimistic when you consider the variety amongst habits. Driving to work, avoiding the cracks in the pavement, thinking about sport, walking the dog, eating a salad, booking a flight to China; they could all be habits, and yet they involve such different areas of our lives. But, to be fair, Maltz didn't invent the 21-day time frame; there are all sorts of stories explaining its origins, most of them standing on science-free ground.

Thanks to recent research, though, we now have some idea of how long common habits really take to form. In a study carried out at University College London, 96 participants were asked to choose an everyday behaviour that they wanted to turn into a habit.[2] They all chose something they didn't already do that could be repeated every day. Many were health-related: people chose things like 'eating a piece of fruit with lunch' and 'running for 15 minutes after dinner'. On each of the 84 days of the study, they logged into a website and reported whether or not they'd carried out the behaviour, as well as how automatic the behaviour had felt. As we'll soon see, acting without thinking, or 'automaticity', is a central component of a habit.

So, here's the big question: how long did it take to form a habit? The simple answer is that, on average, across the participants who provided enough data, it took 66 days until a habit was formed. And, contrary to what's commonly believed, missing a day or two didn't much affect habit formation. The complicated answer is more interesting, though (otherwise this would be a short book). As you might imagine, there was considerable variation in how long habits took to form depending on what people tried to do. People who resolved to drink a glass of water after breakfast were up to maximum automaticity after about 20 days, while those trying to eat a piece of fruit with lunch took at least twice as long to turn it into a habit. The exercise habit proved most tricky with '50 sit-ups after morning coffee', still not a habit after 84 days for one participant. 'Walking for 10 minutes after breakfast', though, was turned into a habit after 50 days for another participant.

The graph shows that this study found a curved relationship between repeating a habit and automaticity. This means that the earlier repetitions produced the greatest gains towards establishing a habit. As time went on these gains were smaller. It's like trying to run up a hill that starts out steep and gradually levels off. At the start you're making great progress upwards, but the closer you get to the peak, the smaller the gains in altitude with each step. For a minority of participants, though, the new habits did not come naturally. Indeed, overall the researchers were surprised by how slowly habits seemed to form. Although the study only cov-

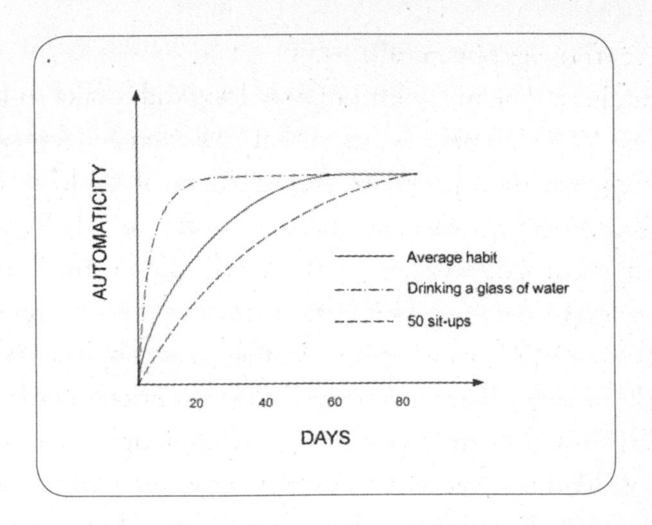

On average, habit formation took 66 days. Drinking a glass of water reached maximum automaticity after 20 days; for 50 sit-ups, it took longer than the 84 days of the study.

ered 84 days, by extrapolating the curves it turned out that some of the habits could have taken around 254 days to form – the better part of a year!

What this research suggests is that taking 21 days to form a habit is probably right, as long as all you want to do is drink a glass of water after breakfast. Anything harder is likely to take longer to become a quite strong habit, and, in the case of some activities, much longer. Dr Maltz and his cheerleaders weren't even close, and all those books promising habit change in only a few weeks are grossly optimistic. Of course, this study opens up a whole new set of questions. The participants were only trying to adopt new habits; what about our existing habits? How much better might they have done using tried-and-tested psychological techniques? And this study doesn't really tell us what a habit feels like, how we experience it, or where it tends to happen.

~~~~~~~~~

What do we actually do all day long? Some busy days slip by in a flash and we remember little. Whether at work or idling around at home, it would be fascinating to know exactly how our time is spent and which parts of it are habitual. Unfortunately, there's a very good reason why we tend to be awful at recalling habitual behaviour, which is to do with its automaticity. So psychologists use diary studies, which give a much more accurate picture of what people are up to than we can get from memory. In one study led by habit researcher Wendy Wood, 70 undergraduates at Texas A&M University were given a watch alarm.[3] Every hour while they were awake, it reminded them to write down what they were doing, thinking, and feeling, right at that very moment. The idea was not just to build up a list of activities, but to see the context in which they occurred. Across two separate studies, the researchers found that somewhere between one-third and half the time people were engaged in behaviours which were rated as habitual. This suggests that as much as half the time we're awake, we're performing a habit of one kind or another. Even this high figure may well be underestimated, since it's based only on young people whose habits haven't had much of a chance to become ingrained.[4]
~~~~~~~~~

So, what were participants in Wood's research up to? Since they were students, the largest category was studying. This included attending lessons, reading, and going to the library, which made up 32% of the diary entries. Amongst these activities, about one-third were classified as habitual. The next category was entertainment, which participants were engaged in for 14% of the time. This included things like watching TV, using the Internet, and listening to music. This time, the percentage of habitual activities went up to 54%. Next on the list were social interactions, which made up 10% of the entries and 47% of which were classified as habitual behaviours. The category in which the behaviours were least habitual was cleaning, down at only 21%, while the category which was most habitual was going to sleep and waking up at 81% (at least they weren't hiding their lazy, slovenly ways!).

More important than precisely what they were doing (especially for those of us who aren't students), are the characteristics of habits. What does it feel like? What's going on in our minds? What emerged from this study, as it has from others, are three main characteristics of a habit. The first is that we're only vaguely aware of performing them, like when you drive to work and don't notice the traffic lights. You know that some part of your mind was attending to them, along with other road-users and the speed limit, but often you can't specifically remember doing so. In Wood's study, participants reported exactly this vagueness about their habitual behaviour. While they were relaxing, watching TV, or brushing their teeth, they reported thinking about what they were doing only 40% of the time. It's one of the major benefits of a habit: it allows us to zone out and think about something else, like planning a weekend trip. Habits allow the conscious part of our minds to go a-wandering while our unconscious gets on with those tedious repetitious behaviours. Habits help protect us from 'decision fatigue': the fact that the mere act of making decisions depletes our mental energy. Whatever can be done automatically frees up our processing power for other thoughts.

A habit doesn't just fly under the radar cognitively; it also does so emotionally. And this is the second characteristic that emerged: the act of performing a habit is curiously emotionless. The reason is that habits, through their repetition, lose their emotional flavour. Like anything in life, as we become habituated our emotional response lessens. The emotion researcher Nico Frijda classifies this as one of the laws of emotion, and it applies to both pleasure and pain.[5] Activities we once considered painful, like getting up early to go to work, become less so with repetition. On the other hand, activities which excite or give us pleasure initially, like sex, beer, or listening to Beethoven's Seventh, soon become mundane. Of course, we fight against the leaking away of pleasure, sometimes with success, by seeking variety. This is why some people feel they have to keep pushing the boundaries of experience just to get the same high.

None of this means we don't feel emotion while performing a habit, it's just that the feelings we experience usually have less to do with the habit and more to do with where our minds have wandered off to. Wood's research found this exact pattern in participants' reports of their emotional experience. Compared with non-habitual behaviours, when people were performing habits their emotions tended not to change. In addition, the emotions that people did experience were less likely to be related to what they were doing than when their activities were non-habitual. The fact that habitual behaviour doesn't stir up strong emotions is one of its advantages. Participants in this study felt more in control and less stressed while performing habits than they did enacting non-habitual behaviours. The moment participants switched to non-habitual behaviours, their stress level increased.

The third important characteristic of a habit is so obvious that often we don't notice it. Perhaps this is partly a result of the automatic nature of habits. Take some typical daily routines: You get up in the morning, go to the bathroom, and have a shower... Later you're in the car when you turn on your favourite radio station... Then, at the coffee shop, you order a blueberry muffin... The connection is context. We tend to do the same things in the same circumstances. Indeed, it's partly this correspondence between the situation and behaviour that causes habits to form in the first place.

The idea that we create associations between our environment and certain behaviours was memorably demonstrated by the Russian physiologist Ivan Pavlov. In Pavlov's most famous research, carried out on dogs, he created an association between being fed and the sound of a bell ringing. Then, after a while, he tried ringing the bell without feeding the dog. He noticed that the dog began to salivate anyway. The toilet, car, and coffee shop are like Pavlov's bell, unconsciously reminding us of long-standing patterns of behaviour, which we then enact again, in exactly the same way as before. This is backed up by research on humans that shows that people tend to perform the same actions in the same contexts. In the diary study described above, most of the behaviours, like socializing, washing, and reading, were carried out in the same place.

It becomes clear just how much context is important for habit whenever you move house or get a new job. Once in a new home, it's suddenly difficult to do the simplest of jobs. Making a sandwich becomes an ordeal as you have to think consciously about where the knives and plates are. It's not just simple tasks that become more difficult; it's all your usual routines. From getting up in the morning to going to bed at night, so many tasks feel like they're being done for the first time. You may even find yourself trying to carry out your old habits in your new home, to no avail: because everything has moved, suddenly those ingrained ways of behaving fail you. The same goes for new jobs. Where once you glided around the workplace on autopilot from one task to the next, in the new job you feel like a fish out of water.

Psychologists have seen how important context is during research on how people cope with changes to their environment. In one study, students' habits were tracked as they transferred to a new university.[6] They were asked how often they watched TV, read the paper, and exercised both before the move and afterwards. They were also asked about the context in which these habitual behaviours were performed. How did they perceive the context, where were they physically, and who was with them at the time? The answers to these questions built a picture of whether the context had really changed with the move from one location to another. For example, it's possible that although a physical location changes, the overall context doesn't. Like hotel rooms, one hall of residence can look much like another; so it might not *feel* that things have changed much.

What the participants reported as they moved from one university to another was that context was important in habit change. They found that if they wanted to cut down their TV and increase their exercise, it was easier to do so after the move. This is because new surroundings don't have all the familiar cues to our old habits. Without these cues, our autopilot doesn't run so smoothly and our conscious mind keeps asking us what to do. That's why moving house is like going on holiday: without your established routines, you have to keep consciously thinking about what you're going to do now. The same thing happened to these students. Instead of automatically watching TV or reading the newspaper, they were more likely to think, 'What did I *plan* to do today?' and 'What do I actually *want* to do now?' As a consequence, a world of possibility opens up.

The rather bland word 'context' can also include other people. Whether we notice it or not, we are heavily influenced by those around us. The researchers in this study found that participants' behaviour was disrupted by any changes in the behaviour of those around them. For example, students reported that they changed their newspaper reading habits if those around them changed theirs. It isn't necessarily the case that we copy other people, just that they tend to cause some change in us. This ties in with the finding that people who live alone report more of their daily behaviours as being habitual than those who live with others.[7] Other people, then, disrupt our routines, sometimes for better, sometimes for worse.

Now we've seen how habits are born, what they feel like, and how much of our daily lives they take up. Three characteristics have emerged: firstly, we perform habits automatically without much conscious deliberation. Secondly, habitual behaviours provoke little emotional response by themselves. Thirdly, habits are strongly rooted in the situations in which they occur. We also know that they can vary considerably in how long they take to form. But how much control do we have over our habits? If we want to make a change, how easy will it be?

2

HABIT VERSUS INTENTION: AN UNFAIR FIGHT

We like to think that our habits follow our intentions. If I want to form a habit, I should be able to. Say I decide to switch from white to wholewheat bread. I buy it from the supermarket a few weeks in a row; I like it so I keep getting it. With each repetition, the habit gets a little stronger, and after a few months I'm picking it up off the shelf without even thinking about it. I intended to eat more healthily, and now I am. Just the same sort of process, with our intentions flowing into our habits, goes on in all sorts of areas of life: learning to ride a bike, dance, or cook. Individual physical actions are built up over time into chains of behaviour we perform automatically.

Mental habits can be built up in just the same way, again

with intentions flowing into habitual ways of thinking. You might decide you're being too harsh on a friend, say, by always thinking they are selfish. You make a mental note to spot a more benevolent trend in their behaviour. You notice when they buy you a drink and listen to your problems. Small things, but steps in the right direction. Sure enough, you start to think of them as less selfish. Unconsciously, the habitual ways in which you think about your friend have changed.

Our mental habits can change in this way because our minds are so good at spotting patterns; indeed, it's one of the mind's chief functions. Our ability to spot patterns at low levels and build them up into a habit, based on our conscious intentions, enables us to reach much more complex goals. Here's an example from a classic psychology study. Participants sat in front of a computer for almost an hour, pressing one of four buttons corresponding to where a cross appeared on the screen.[1] Naturally, it was very boring, but the designers of the experiment had a little trick up their sleeve. Unknown to the participants, there was a pattern in where the crosses appeared. Despite it being consciously undetectable, the participants began to respond faster as the study went on – they were learning the pattern. When interviewed afterwards, though, none had noticed anything: they had learnt it without realizing. This is a study about unconscious learning, but it demonstrates how mental habits can grow out of patterns. Here, an unconscious learning process was evolving in the service of a higher-level intention: to do well on the test and please the experiment's designers.

When you learn to serve in tennis or reverse a car into a tight space, it's the physical equivalent of this unconscious mental learning process. Lots of small unconscious actions are built up to achieve one big conscious goal: taking a serve or parking a car. In the mental realm, mathematics is an early example of this building-up process. At school, we learn a series of operations we can perform on numbers to reach a goal: say, working out the average height of our schoolfriends. Although learning these basic operations (addition and division) can be excruciating for young minds, they soon become second nature. Later on, we can perform them almost without conscious thought, which enables us to complete much more complex calculations. Once again, the habit of particular mental or physical operations helps us achieve a whole series of higher-order goals.

We all have an intuitive sense that our habits are built up purely in the service of our goals (remember that bad habits are also goal-oriented, although the goal may not be a good one, like getting drunk to forget one's problems). Indeed, the stronger people's habits, the more they believe that those habits are goal-oriented.[2]

Our intuitive sense that intentions lead straight into habits is far from just a lay understanding. Many influential psychologists have expressed exactly the same idea. Generations of first-year undergraduate psychologists are taught that intentions are a major key to predicting behaviour. They learn theories with grand-sounding names like the 'model of interpersonal behaviour'[3], the 'theory of planned behaviour'[4], and the 'theory of reasoned action'[5], which all suggest that when we form an intention, it leads us to act in line with that intention. These are influential ideas across different sub-disciplines of psychology and they underpin much research.

Now these theories are being challenged because, like our intuitive understanding, they don't tell the full story. We may like to think our intentions flow directly into our habits, but often they don't. It's an idea we resist because it strikes at our sense of having free will. We like to think that things happen for a reason, and that one of those reasons is because we decided it would happen, or at the very least, that someone else decided it would happen. Yet habits don't flow solely from our intentions, and there are studies that demonstrate this.

Worse for our sense of agency, it's possible for intention and habit to be completely reversed. Sometimes we unconsciously infer our intentions from our habits. How the habit started in the first place could be a complete accident, but we can then work out our intentions from our behaviour, as long as there's no strong reason for that behaviour. Say I take a walk around the park every afternoon and each time I follow a particular route which takes me past a duck pond. When asked why I take this route, I might reply that I like to watch people feeding the ducks. In reality, I just walked that way the first time, completely at random, and saw no reason not to do the same the next day. Now, after the habit is established, I try to come up with a reason, and the ducks spring to mind. I end up inferring intention from what was essentially just chance.

We know people regularly do this sort of backwards thinking, and really believe it. One of the most famous examples in psychological research is cognitive dissonance. This is the idea that people don't like to hold two inconsistent ideas to be true at the same time. Studies conducted more than half a century ago find that when people are induced into behaviour that is inconsistent with their beliefs, they simply change their beliefs to match.[6] It's like when someone ends up spending too much on a new car. Instead of feeling bad about the clash between their original plan and what they've actually done, they prefer to convince themselves that the car is worth the extra money. This is a result of our natural desire to maintain consistency between our thoughts and actions. We all want to be right, and one thing we should all be able to be right about is ourselves. Backwards thinking allows us to do just that.

But surely we would know if we were doing this kind of backwards thinking? Unfortunately, though, we have little access to these sorts of unconscious processes. It turns out that in experiment after experiment, psychologists can change minds without participants realizing. In one study on attitudes, people clearly changed their mind on an issue after being bombarded with reasons to do so.[7] Despite this, they claimed the arguments had had no effect on them; indeed, they thought their new attitudes were what they had always thought. It seems politicians aren't alone in blanking out their U-turns. Like it or not, we're all capable of it.

What we've explored so far are two extremes: when we create habits intentionally for a particular purpose, and when we infer intentions from our behaviour. In real life, though, both of these processes happen at the same time, and habit is a combination of our intentions and our past behaviour. So here's the crucial question: what kind of combination? Can the intention to start eating healthily or get a new job really overcome the habit of eating junk food and going to the same office every day?

We already know quite a lot about this question because psychologists are very keen to change people's behaviour, hopefully for the better. Studies on donating blood, exercising, recycling, and voting have all examined whether it's possible to change people's habits. One of these tested if participants could predict their own consumption of fast food, how much they watched TV news, and how often they took the bus over a week.[8] Each person was asked how much they intended to carry out each of those three behaviours over the coming week. Then, they were asked how often they had performed each behaviour in the past. These are the measures of intention and habit. Over the next 7 days, participants noted down how often they went into a fast-food restaurant, watched TV news, and took the bus.

The results showed that when established habits were weak, intentions tended to predict behaviour. So, if you don't watch TV news that much, your intention for the coming week, whether it's to watch more, less, or the same, is likely to be accurate. Good news for our sense of self-control. Here comes the bad news. As habits get stronger, our intentions predict our behaviour less and less. So, when you're in the habit of visiting fast-food restaurants, for example, it doesn't matter much whether you intend to cut down or not, chances are that your habit will continue.

It gets worse, though. Participants were also asked how confident they were in predicting their behaviour over the coming 7 days. An unusual result emerged. Those with the strongest habits, who were the least successful in predicting their behaviour over the coming week, were the most confident in their predictions. The finding is striking because it hints at one of the dark sides of habits. When we perform an action repeatedly, its familiarity seems to bleed back into our judgements about that behaviour. We end up feeling we have more control over precisely the behaviours that, in reality, we have the *least* control over. It's another example of our thought processes working in the opposite way to our intuitive expectations.

~~~~~~~~~

Considering how powerful habits are in the face of conscious intentions, it is vital to know what a strong habit is compared with a weak habit. For example, is buying a pair of shoes once a month a habit? What about reading the newspaper every day, or attending a community meeting twice a year? How often before we find it increasingly difficult to stop ourselves or, put the other way round, no longer have to force ourselves? Psychologists have looked at this in a review of 60 different research reports on habitual behaviour.[9] They classified habits into two categories. In the
~~~~~~~~~

first category they put things like exercising, coffee-drinking, and using a seat belt; the kinds of things that you might do at least once a week. In the second category they put the kinds of things we might only do a few times a year. They included things like donating blood or getting a flu jab, but could just as easily include going to the dentist or getting a haircut. The other important thing they took into account was the context in which each repeated action took place. Context is a vital component of habitual behaviour because we tend to perform the same actions in response to particular situations.

Across all the studies, intentions emerged as the strongest predictor of future behaviour. Overall, people were doing what they intended. Yet when habits were divided up into either those performed about weekly or those performed approximately yearly, a big difference emerged. Once again, when behaviours were performed weekly, established habits tended to rule people's behaviour in comparison to any plans they'd formed to act differently. It was only when behaviours were performed only once or a few times a year, like getting flu jabs or donating blood, that intentions took over from autopilot. Once again, the situation was also important since habitual behaviours performed in stable situations – like always ordering a latte in a coffee shop – are even less susceptible to our intentions.

This suggests that the difference between a strong and a weak habit is somewhere in the region of whether it is performed weekly or only a few times a year. This means strong habits could encompass an enormous amount of our behaviour. If you think about the things you might do on a weekly basis in the same context – say, visiting a restaurant or watching a film – it feels as if these decisions are highly intentional. But the research would suggest these types of behaviours are close in nature to daily actions such as wearing a seat belt, catching up on the news, or checking your email. We have less intentional, conscious control over these types of behaviours than we would like to think.

For years, psychologists have tried changing people's bad habits by targeting their intentions. Hundreds of studies have attempted to help people adopt a low-fat diet, do more exercise, wear a bicycle helmet, use a condom, take a college course, quit smoking, put on sunscreen, and many, many more laudable causes. When the results are added up together, they don't look too clever. One review of 47 of the most rigorous of these studies produced sobering reading.[10] On the positive side, psychologists are very successful at getting people to change their goals and intentions. After various psychological techniques have been used on them, people in these studies definitely want and intend to change. Unfortunately, the problem comes with breaking down existing habits. Although people *intend* to change, when habits are strong, actual behaviour change is relatively low.

Despite all this talk of how weak intentions are in the face of habits, it's worth emphasizing that much of the time even our strong habits do follow our intentions. We are mostly doing what we intend to do, even though it's happening automatically. When washing our face each day, getting an espresso on the way to work, or cleaning our glasses, it's because at some point in the past we consciously decided (or someone decided for us) that these things were worthwhile activities, so we kept repeating them until they were automatic. This probably goes for many habits: although we perform them without bringing the intention to consciousness, the habits still line up with our original intentions. Even better, our automatic, unconscious habits can keep us safe even when our conscious mind is distracted. We look both ways before crossing the road despite reminiscing about a rather depressing holiday we took in Brazil, and we put oven gloves on before reaching into the oven despite being preoccupied about whether the Brussels sprouts are overcooked. In both cases, our goal of keeping ourselves alive and unburnt is served by our automatic, unconscious habits. It's only for the minority of bad habits that we want to change that things become tricky.

There's no doubt that there are plenty of occasions when we can successfully make or break our habits. Still, what we find from the research on habits and intention is that our conscious decisions aren't as strong as we'd like to think. In some ways, this is a comforting thought. It means that all those times we tried to change our behaviour and failed because old habits intervened, there was a good reason: the sheer power of strong habits. Studies show that it's normal for strong habits to override our conscious intentions. Combine that with how long habits take to form, and it's no wonder we find our everyday behaviours difficult to change.

So why do our habits not submit to our intentions? To answer that question, we have to take a trip into the deep, dark, mysterious world of the unconscious, where the secrets of how our habits operate are buried.

3

YOUR SECRET AUTOPILOT

Imagine you're at a friend's party. There are quite a few new faces, so you're scanning the room. Then your gaze lands on an attractive stranger on the other side of the room. You look away, you look back. The first hint of a smile plays across their lips. Suddenly, you're nervous, your mind goes blank, you want to go over and you want to run away, both at the same time. You turn around too fast, bump into someone, almost spilling your drink. Then you take a few deep breaths, compose yourself, and pretend to be looking around for someone you know while you try to track down the attractive stranger. There they are, over there, half concealed by a lamp. A friend taps you on the arm to ask who you're looking at…

Now, let me ask you a question. Do you think you'd be able to describe accurately why you find this person attractive? Indeed, how good are we in general at pinpointing what it is about others that attracts us?

Before you answer, consider a sneaky study carried out by Swedish psychologists.[1] People were shown pairs of female faces on playing-card–sized photos, one in each of the experimenter's hands. They pointed to whichever of the two faces they found more attractive. The experimenter then passed the card to the participants and asked them to describe exactly why they found that face attractive. But this is a psychology experiment, so there's a twist in the tale. Sometimes, when the experimenter passed the card to the participants, there was a little sleight of hand involved. This resulted in half the participants staring at the female face they didn't choose. So half the participants were being asked to justify a decision that, in reality, they hadn't made. A few spotted the trick, but most didn't; they were then asked to describe exactly why they had chosen that face.

Think about what you'd expect to get. If the face was second-best for you, then wouldn't your enthusiasm at least be dampened? Perhaps the information would be processed unconsciously, leading to a subtle difference in your report? For example, we might be more uncertain or more vague about why we preferred this face. But analysing the participants' reports, the researchers couldn't find any difference between the two groups. Both the participants looking at the photo they chose and those looking at the one they didn't seemed sure of their reasons, were equally specific, and expressed equal levels of emotionality. The verbal reports gave no clue that the switch had been performed. The researchers gave this phenomenon a clever new name: choice blindness. This, then, is the idea that under certain circumstances, we are actually oblivious to the choice we have made.

To take you back to the opening scene: how sure are you now that you'd be able to describe accurately what it was you saw in that attractive stranger? Hopefully, if you were sure before, you're slightly less sure now. What this study hints at is the strange nature of the unconscious, which is central to an understanding of how our habits work the way they do and what we can do to change them.

~~~

For thousands of years, humans have been trying to make sense of what's going on in their own minds. One of the most famous voyagers to the inner self was the psychoanalytic pioneer Sigmund Freud. Although nowadays he doesn't enjoy the same scientific prominence he used to, his ideas about the unconscious have taken hold of the popular imagination. So much so that we tend to think it's possible to reach down into our unconscious and find out about ourselves. The process of psychoanalysis, Freud liked to explain, was not unlike that of an archaeological dig. While it can be difficult to unearth the truth about yourself, it is nevertheless there, buried deep down below layers of neuroses and complexes and other strange motivations and desires.

Many modern psychologists take quite a different view
~~~

of the unconscious. It is best articulated by Timothy D. Wilson of the University of Virginia, who has long been interested in what we do (and frequently don't) know about ourselves. Over the years, Wilson and others have sent thousands of participants off on these archaeological digs into the unconscious mind to see what comes up. In one study, researchers set themselves up in a shopping centre pretending to carry out a consumer survey on nightgowns and tights.[2] Passers-by were asked to evaluate what they were told were four different nightgowns and four different pairs of tights. In fact, all four items were identical. Quite by accident, they discovered that people seemed to prefer the item that was on the far right, and this was most pronounced for the tights. The right-most pair, although identical to the left-most, was preferred by a factor of four to one.

But did people notice that it was because they were on the right? Could they dig down and work out what was going on? Apparently not. When asked why they had chosen a particular item, no one mentioned its position. Even when the experimenters suggested that the position might have an effect, most participants looked at best very confused and at worst utterly dismissive. So people didn't have a clue why they preferred one identical pair of tights over another. Score one for the unconscious and zero for the conscious.

Another of Wilson's studies looked at the inverse situation: when people think something will unconsciously influence them when, actually, it doesn't. In this study, participants read a passage from the novel *Rabbit, Run* by

John Updike. The extract from the book involves an emotionally charged scene in which an alcoholic mother, while washing her baby in the bath, accidentally drowns her. The participants were split into four groups, who were presented with different versions of the passage:

1. The scene was presented in its entirety.
2. A part of the scene – a description of the baby's messy crib – was deleted.
3. A different part of the scene – a physical description of the baby – was deleted.
4. Both (2) and (3).

Afterwards, participants rated the emotional impact of whichever passage they had read on a simple scale from 1 to 7. Then, participants in condition 2, 3, or 4 were shown the deleted scenes and asked if it would have made any difference to the emotional impact of the whole extract if they had been included. On average, most of the participants thought the deleted parts would have increased the emotional impact. But when the researchers looked at the ratings, it was clear that the emotional impact was unaffected by deleting either or both of these sections. So here we have people thinking something will unconsciously affect them when, in fact, it made no measurable difference at all. Score two for the unconscious and zero for the conscious.

Now the examples get more personal and even slightly uncomfortable. You might well think it isn't that big a deal that you don't know why you choose particular products, or can't accurately predict the emotional impact of literary works. These are not that important. So let's come a little closer to home. Let's talk about personality, attitudes, and self-esteem. These are three things about ourselves that we should be able to judge accurately.

Once again we find shocking deficits in our self-knowledge. Take the case of shyness. One study compared people's self reporting about their shyness with an implicit test of shyness, that is, by seeing what they do in a real situation, rather than what they say they do.[3] Now, of course, there is some overlap between self-reports and implicit tests: people who are raging extroverts don't report being really shy. But what this study finds is that there is not as much overlap as we would expect. We seem to know something about our own personalities, but not as much as we'd like to think.

Attitudes are great examples of where people say one thing but their actions reveal something else. We all know people who secretly watch TV programmes that they would never publicly admit to liking. The most incendiary example is race, where people claim they're not racist but their behaviour suggests otherwise. It is possible people are trying to keep unsavoury attitudes quiet, but the research suggests that people are actually successfully hiding it from themselves.[4]

Perhaps the most incredible example is self-esteem. Surely we know how high our own self-esteem is? Well, psychologists have used sneaky methods of measuring self-esteem indirectly and then compared them with what we explicitly say. For example, one study put participants through a five-minute interview designed to make them feel that their personality was being probed.[5] They were asked the types of questions that psychologists are stereotypically supposed to ask, such as: 'If you could be any sort of animal, what animal would that be and why?' This was a smoke-screen; in fact, they wanted to see how much nervous body language participants exhibited – this was the real measure of self-esteem. What they found was only a very weak connection between how high they thought their self-esteem was and how much nervous body language they displayed.

It seems almost unbelievable that we aren't aware of how high our own self-esteem is since it's such an integral part of ourselves. Amazingly, some studies find almost no connection at all.[6] It's as if you were to ask someone what colour their eyes were and where they were born, and the best they could do was 'darkish' and 'somewhere in the Northern Hemisphere'. There's even some evidence that the more we try to think about our self-esteem, the less accurate we become.[7] Once again, with self-esteem, as with the other aspects of self-knowledge, we are strangers to ourselves.

~~~

The fact that our unconscious doesn't completely control us is mostly down to our frontal lobes, the part of the brain situated just above our eyes. This area is most associated with all the higher functions such as reasoning, memory, and planning, but it also works to monitor and inhibit our actions. When this area of the brain is damaged, control over habits can be lost.

The French neurologist François Lhermitte was the first
~~~

to document systematically a type of disorder he called *utilization behaviour*.[8] The patients he described all had damage to the frontal lobes of the brain: some had Alzheimer's disease, others had had surgery for cancer and some treatment for aneurysms. What Lhermitte noticed was that many of these patients exhibited a similar type of behaviour. When a pair of glasses was put on the table in front of them, they would reach out, pick them up, and put them on. Nothing odd in that, you might say. Except when another pair of glasses was put on the table, they would pick those up and put them on over the first pair. And with another pair they would repeat the same action. Patients with utilization behaviour show this pattern for all sorts of habitual actions and without any internal motivation. If a glass of water is placed in front of them, they drink it even though they're not thirsty; if food is brought, they start to eat it despite just having had lunch; if a comb is put on the table, they pick it up and use it although their hair is perfectly tidy.

What's even more strange is that they will perform all these actions after being specifically told not to. When asked why they drank from the water despite not being thirsty and being told not to touch the water, they simply reply: 'Because you held out the objects to me and I thought I had to grasp and use them.' Then, sometimes, they would sit there asking themselves, 'Must I use them?' This utilization behaviour only seems to happen when the patients already have an established habit. For those who don't smoke, cigarettes and a lighter provoke no automatic behaviour. But if the experimenter reaches for the pack and takes a cigarette, the patient will light it for them.

At the other extreme are patients who seem only too aware of how their habits are being unconsciously cued. One example is known as 'alien hand syndrome'. Here, patients find their hand performs all kinds of actions that they don't want it to. Like those with utilization behaviour, they will reach out and grab glasses of water, door knobs, or clothing, despite having no conscious desire for water, going through doors, or undressing. Their experience is one of complete detachment from the 'alien hand', as though someone else were operating it. The syndrome was used to great comic effect by Stanley Kubrick in his 1964 film *Dr Strangelove*, in which the eponymous doctor, played by Peter Sellers, can't keep his 'alien hand' under control. Away from the world of Hollywood, though, patients with this problem find it very difficult as they really experience the hand as though it were being externally controlled.

This is a glimpse into two frightening worlds in which unconscious habits take complete control over the physical body; one without the patient realizing, and another, more distressingly, with full conscious awareness. Fortunately for most of us, these extremes of behaviour are only something we can try to understand from the outside. Although they're severe examples, they do demonstrate how our habits are continually bubbling up from the unconscious. When we see glasses of water, door knobs, or plates of food, somewhere deep in our unconscious, automatic processes are being initiated. The fact that we don't always perform these habits is down to other inhibitory processes, which try to stop us eating and drinking when we're not hungry or thirsty and opening doors when we don't want to leave rooms.

~~~~~~~~

The unconscious mind is carrying out all sorts of high-level thinking which we don't have access to, try as we might. This includes basic perceptual and motor processes that we'd expect, such as how to catch a ball, recognize the face of a loved one, or reverse a car into a tight space. Generally, it's not too worrying that there are things our bodies know that our conscious minds don't. Less intuitively, and less happily for our own sense of self, we find that we have little access to the kinds of thinking processes that should be transparent, such as our own attitudes, our personality, and our self-esteem.

Over the years, and over many hundreds of studies, a new view of the unconscious has emerged, a view that diverges from Freud's theories in a central way. Freud thought we could dig down through the archaeological layers to get at the truth of why we think and do the things we do; many modern psychologists think otherwise. Rather than a series of archaeological layers which can be carefully scraped away, the centre of our unconscious is more like the Earth's core: we get the results in the form of emotional earthquakes, thought eruptions and the rest, but the actual causes can be extremely mysterious. That's why sometimes our emotions, attitudes, and decisions can fluctuate for no reason that's accessible to our conscious selves, leaving us
~~~~~~~~

floundering.

None of this stops us from trying to guess what's going on down there, which we do all the time, and researchers have been fascinated to see what happens when we try. The results are not heartening. In one study, students waiting in the queue at a university dining hall filled out a questionnaire asking them why they liked their chosen drink.[9] Others just made their choice without thinking twice about their preference – they acted as a control group. Then the researchers looked in the cups to see how much of each drink they had drunk. What emerged was that participants who had really thought about how much they liked their chosen drink were less accurate in predicting how much they would actually drink than those who hadn't thought about it. In other words, thinking carefully about their preferences lowered people's ability to predict successfully their own behaviour.

In another study, participants were given a choice between two types of poster to take home: an artistic one or a humorous one.[10] While some were asked to think about the reasons for their choice, others just chose. An interesting thing happened: trying to work out why they liked the poster made participants more likely to choose the humorous one rather than the artistic one. Then they took them home and reported back after a few weeks about how satisfied they were with their choice. It emerged that those who chose the humorous poster were less satisfied. When the researchers looked more closely at the reasons, they noticed that, on average, people found it easier to come up with reasons to like the humorous poster and, at the same time, to say why they didn't like the art poster. So because the art poster was more difficult to think about, people chose the humorous poster. But when they got the humorous poster home, it didn't seem quite as funny. What we're seeing here is that people's choices are affected by what thoughts arrive most easily in consciousness, not necessarily what's in the unconscious. On top of this, thinking too hard about the reasons for our decisions can make us less happy with those decisions.

This doesn't necessarily mean that introspection *always* makes us worse predictors of our own behaviour or less happy with our choices, but it certainly shows the potential dangers. Still, the typical story of our interaction with our unconscious is frustrating in the extreme. Because we don't have direct access to the reasons we do things, we make up some reasons based on our own personal preferences, theories about the world, and any other available conscious information we can lay our hands on. It's like Aesop's fable about the fox that spots some grapes high up on a branch. The fox tries to jump up and get them, but finds they are out of reach. In a flash, the fox performs a mental U-turn and decides it doesn't want the grapes any more because they're probably sour. With this U-turn, the fox is protecting itself from the frustration of not being able to get the grapes (incidentally, this is where we get the phrase 'sour grapes'). Smokers are doing something similar when they tell you that they know someone who smoked forty a day and lived until 100 years of age, or that if the smoking doesn't kill them, something else will. These are rationalizations of the kind our unconscious is spinning all the time, but without our knowledge.

~~~~~~~~

The fact that the unconscious is almost impossible to penetrate looks like a problem for anyone who wants to change their habits, since they live mostly in the unconscious part of the mind. Really, though, awareness of the power of the unconscious to guide and change our thinking and behaviour is the first step to change. If we deny how much of our thought and behaviour is unconscious, we'll have less chance of making changes stick. Probing the unconscious to try to explain our habits is a waste of time – indeed, it may even be counterproductive – but becoming more aware of our behaviour, something we can notice, is very helpful.

This is because the habit itself is one of the most important clues as to what is going on in the unconscious. We can use our memories and conscious awareness to piece together a picture of what might be going on down there, at our core. With these clues, and an insight into how they are produced by the interaction between what habits we want and what habits we actually get, we can take better control of ourselves. And it's to this interaction that we turn in the next chapter.
~~~~~~~~

4

DON'T THINK, JUST DO

The other day I was halfway through watching a film on TV when the power went out. Our wiring isn't the best, so I looked out of the window to check that everyone else's power was out. It was. Then I called the electricity company to complain and was told that engineers had been dispatched. It was dark, so I stumbled about the house using the glow from my mobile-phone display. Having found the candles and matches, the flat was soon lit by a soft glow. I'm irritated and confused because I was totally absorbed in the film, and suddenly I've been plunged back two centuries into a time before electricity and Hollywood.

While I was waiting for the power to be restored, I went to the lavatory. As I walked in, my hand went up to the light switch and clicked it on. For a fraction of a second, I stood there confused, wondering why it hadn't had its customary effect, then I snorted at my own stupidity. Not only have I looked out of the window to check that the power cut covers my area, I've phoned the power company and lit candles, and yet I was still trying to turn on the light. There's worse to follow when, two hours later, the power still hasn't been restored. I went into the bathroom again, and my hand reached up and clicked the switch again. I stood there dumb-struck, wondering what could possibly have been going through my mind. The answer is: absolutely nothing.

This sort of behaviour is the kind of thing the famous American behavioural psychologist B. F. Skinner would have found comforting, because it fitted perfectly with his view of habits. The way I acted with my light switch looks similar to how his pigeons behaved in his most famous experiments. In one of these, Skinner put hungry pigeons in a box and fed them once every 15 seconds. Soon, they began to exhibit unusual behaviours. One pigeon began stretching its neck just before the food was delivered; another started walking in circles; yet another stuck its head in the corner. Skinner's explanation was that the pigeons learnt to associate moving their necks, or walking in circles, or sticking their heads in the corner, with the reward of food. So they came to 'think' that these movements somehow caused the food to arrive. The pigeons had become superstitious.

The classic view of habits sees humans as much more complicated versions of Skinner's pecking pigeons. In this view, our habits are built up quite simply in response to rewards from the environment. For example, my job pays me money, which I want, and I learn from the environment that if I work harder, I can get promoted and get more money. So I develop a work ethic. Or, I want to be liked more and I notice that smiling helps, so I start smiling more. And my habit of smiling at people is born...and so on. All the time, I'm noticing (whether consciously or unconsciously) that a certain type of behaviour is rewarded, and the more it is rewarded, the more I perform the behaviour in the same situation. These are simplistic examples, but if you imagine them built up layer on layer, the idea is that you can get a view of how humans acquire their habits.

To return to the pigeon for a moment: it wasn't really superstitious; it's too stupid for such a complicated idea. Whatever you say about me, I'm pretty sure I can outwit a pigeon, so how come I'm behaving just like it and trying to turn on a light switch during a power cut? It's because, to some extent, our responses to the world are quite pigeon-like, but clearly, to reduce all human behaviour to that of pigeons is ridiculous.

One of many ways in which humans differ from pigeons is that they have dreams. We can't definitively say that a pigeon doesn't dream of one day owning its own statue to defecate on, setting up a colony on Mars, or being supreme leader, but it seems highly unlikely. In contrast, humans are chock-full of dreams. To try to reach our dreams, we have all kinds of goals, many operating all at the same time, waiting for the right opportunity to be activated. We dream of a clean house, a well-educated son or daughter, and a promotion at work, so when the time is right we get out the mop and bucket, or research the right schools, or practise our brown-nosing. But when, exactly, is the right time?

Certainly, many of our habits are cued at exactly the right time. Like pigeons, we have learnt when to execute particularly complex behaviours in response to the right situations and rewards. We spy on other diners in restaurants to gossip about them, browse Facebook at work for entertaining pictures, and ring a good friend when we need cheering up. All perfectly normal, everyday routines. What's more interesting, though, is when our habits go off the rails, when they don't line up with our goals, when the pigeon inside all of us goes haywire. We want to lose weight, but we keep eating mountains of cake; we dream of a promotion, but end up procrastinating at work; we want to reduce our drinking, but end up ordering another bottle of champagne. Here, our behaviour seems to be in direct contravention of our goals. What we want isn't what we get. Part of the explanation is that habits can be performed unconsciously and strong habits are difficult to change, so this takes us part of the way there. But, in this chapter, we take it a step further to show why a pigeon theory of mind can't explain how we perform habits.

~~~

For many of our everyday activities, our habits serve us very well, from getting dressed to looking both ways before crossing the road, to asking after each other's health. The fact that we perform them unconsciously is more than just
~~~

handy; in fact, ordinary life would be impossible if they weren't. But sometimes habits can be cued up by the environment with little or no reference to our goals or intentions.

Take some rather crafty research into habits of thought headed up by the social psychologist John Bargh. Participants were split into two groups and a little trick was played on them.[1] They were asked to unscramble five words and make a four-word sentence. For example, they were given things like: 'he it hides finds instantly'. It doesn't take too much imagination to discard the word 'hides' and come up with 'He finds it instantly'. For half the participants the sentences were just to keep them busy, but for the other half there was a secret message. The sentences had lots of words which are stereotypically associated with old people; here are a few: 'old, lonely, grey, selfishly, careful, sentimental, wise, stubborn, courteous'. Apologies to more mature readers, but this test is designed to elicit stereotypes, so it has to be crude.

After they had finished the test and thought the study was over, that's when it really got going. A confederate of the researchers sat on a nearby seat to see how long it would take each participant to cover the 9.75 metres to a strip of tape set up as a surreptitious finishing line. Without knowing it, the participants were involved in a race, and the results showed that the losers were those who'd been fed the old-related words. On average, they took a full second longer to cover the distance (8.3 seconds) than those who hadn't had the stereotype activated (7.3 seconds). What had happened in people's minds was that they were reminded about the idea of being old. Because we have habitual ways of thinking about old people – what we usually call stereotypes – it's easy for these ideas to be activated unconsciously. Then, we act in line with these stereotypes without even realizing it. Indeed, only one person in this study noticed that lots of the words were related to a stereotypical view of old age. What this study demonstrates is that a habit of thought activated outside conscious awareness can measurably change people's behaviour.

If that example strikes you as a little depressing, then you'll be happy to hear that you can improve people's performance – effectively get them to perform good habits – by just the same method. In one piece of research, Asian-American participants were invited to take a maths test.[2] Before they did it, though, some were primed with words that would activate stereotypes about Asian people, namely a supposed superiority at maths. This was done by flashing up words on a screen for less than a tenth of a second: this is too quick to perceive consciously, but slow enough for the unconscious to register (it's the old subliminal advertising trick, which was actually a hoax originally, but does work). The words flashed up were things like 'Wok', 'Asia', 'Chinatown', and 'Hong Kong'. The other half got words that had no relevance to ethnic stereotypes. Again, apologies to Asian people; the unconscious isn't as politically correct as it should be.

What they found was that the priming had quite a marked effect on participants' performance. Asian-Americans who had been primed with the stereotype got almost twice as many of the questions right as the other group. That is a serious performance gain for an unconscious cue. When the researchers looked closely at the data, they saw the reason for the performance boost. After being subliminally primed with an Asian stereotype, Asian-Americans attempted more questions. It seemed that being reminded of the stereotype made them try harder. A habit of thought cued a habit of persistence.

Now on to an important question: can all this unconscious priming of good habits of thought make you any money? For example, could it help you win *Who Wants to be a Millionaire*? Perhaps it could. In one study, participants were primed with either the idea of 'intelligence' or that of 'stupidity'.[3] Then they were asked a number of general knowledge questions like 'Who painted *La Guernica*?' (Answer choices: a. Dali, b. Miro, c. Picasso, d. Velasquez.) Just the same effect as before emerged. Subliminally priming people with the idea of intelligence meant they were better able to pluck the correct answer from memory (it is Picasso).

These studies show that habits of thought and behaviour can be activated automatically by people and things around us. We are continually being bombarded by subtle – and sometimes none too subtle – cues about how to behave. We process these automatically and unconsciously, and these impulses emerge as our habits, which we start performing without conscious thought. It's an extension of what we saw in the earlier studies by Wendy Wood. Students who moved from one university to another tended to change their habits because their environments changed. They weren't seeing the same people or being exposed to the same cues so their TV watching habits, exercising habits, and so on changed. With fewer habits being activated, they became more responsive to their own intentions.

What these examples demonstrate is 'direct cuing', that is, there's a direct link between some aspect of the environment (in this case, living somewhere different) and the particular behaviour (watching less TV).[4] But it's also possible for habits to be cued up in more roundabout ways. It's here that we enter the realm of 'motivated cuing'. This is an odd effect where habits can become completely divorced from the goals they were originally designed to accomplish. This is a world where our ability to see why we're behaving in particular ways becomes even more blurred.

Let's imagine for a moment you're a student at university. You're free of the constraints of home and family. You've escaped your parents, but you're not yet caught in the standard routines of adult life: you're not married, don't have children, don't have house and car payments to make. You're as free as you'll ever be; you like to socialize and, like many students, you do so while drinking alcohol. With a pint of beer in your hand, you can enjoy your new-found freedom, your new friends, and a future that seems endless. It's a heady concoction, and the feeling of being around your friends, of enjoying each other's company, is just as intoxicating as the alcohol. You're not drinking to escape, as adults sometimes do, from the boredom of their routines, nor are you trying to dull your senses to enable you to sleep. Quite the reverse, you are tasting freedom: the first freedom of adulthood and the promise of more to come.

Psychologically, though, what's happening is that you are learning an association between the habit of drinking alcohol and the pleasure you get, not just from the feeling of intoxication, but also from the pleasures of socializing. In fact, for many 'drinking' societies around the world, which include the US, the UK, and other countries, this is a link that's ingrained by culture. In some ways, what we really want is just to socialize with other people, but because of established rituals we end up drinking alcohol at the same time.

When you think about it, this is an odd claim, because it is saying that drinking alcohol serves almost no purpose – you might as well perform any kind of ritual, like weaving a basket, dancing, or intermittently breaking into song. Your real aim is to get pleasure from socializing, and the drinking is merely a by-product. Many drinkers, occasional or otherwise, would argue that the alcohol enhances the experience or even enables it, rather than being a by-product of it. There is certainly some truth to these points, but there is still a significant element of habit in there.

Just this disconnect between goal and habit in drinking behaviour has been shown in the laboratory. Psychologists have used jumbled sentence priming techniques similar to the slow-walking research described above.[5] Again, the idea is to prime the unconscious with ideas that the conscious mind is not aware of, and then see how people's behaviour changes. In one experiment, instead of focusing on stereotypes or intelligence, though, they wanted to manipulate the desire to socialize. They did this by asking half their participants to think about cities that are good for socializing and half to think about cities that are good for historical sites. The idea was to have half of them unconsciously thinking about the goal of socializing, with the other half acting as the control group. Then, as a thank you for taking part in the study, participants chose a discount voucher for either tea/coffee or beer/wine. What the researchers found was that for those who were habitual drinkers, unconsciously thinking about socializing made them more likely to choose the alcoholic drink. For those who weren't habitual drinkers, though, it didn't make any difference. Just the mere idea of socializing was activating the idea of drinking alcohol. This shows a disconnect between the goal (feeling good from socializing) and the method being used to reach that goal (drinking alcohol).

One of the ironies of the effect of alcohol is that it reduces our ability to reason effectively. This means we have to rely on our habits *more*. So once you've had a drink, it's even easier for the goal of feeling good about socializing automatically to activate the habit of more drinking. And we all know where that leads…

What's going on in this study is similar to what happened to me when I tried to turn on the bathroom light during a power cut. My goal was to illuminate the room, which was cued up by my entering the bathroom and finding myself standing in darkness. And so it cued up a habitual behaviour: turning on the light switch. Unfortunately, the habit was useless under these circumstances, but because the association is so strong I perform the behaviour anyway, despite the fact that it doesn't achieve my goal. In this case, the error of my ways is soon obvious as I'm still standing in darkness; in contrast, dangerous habits, like drinking too much alcohol, creep up on people in a much more insidious fashion.

Although we don't realize it, these disconnects between our everyday goals and the habits we perform are going on all the time. Because we've built up such strong associations between habits and goals, we don't notice when those habits stop achieving our goals because real life is more complicated than an on–off switch. Let's say you want to change your travel habits. You decide that, every now and then, instead of taking the car for short journeys, you'll walk. The shop is only a 15-minute walk away, so why not get a little exercise? One day you run out of milk, and before you know it, you're sitting in the car with the keys in your hand. Why is that, and what happened to the goal of habit change?

Dutch psychologists Henk Aarts and Ap Dijksterhuis have looked at a very similar situation in research on travel habits in the Netherlands.[6] They found that those people in their studies who were habitual bicycle-riders automatically thought about their bikes when primed with the idea of travelling. This didn't happen to people who weren't habitual bike-riders. This is precisely the link that we are fighting against when trying to change habits. Because of the automatic, unconscious association between a goal (going to the shops) and habit (taking the car), we're already in the car and halfway to the shops before we think anything of it.

What's emerging is a more subtle view of habits than a pigeon theory can support. There isn't just a simple connection in our minds between something we want (like sex, money, or chocolate) and some behaviour we perform to get it (like Internet dating, robbery, or using a vending machine). Unlike pigeons, we have plans, goals, and dreams, as well as desires and drives: it's why human life is so complicated. The catch is that our goals and desires can be activated unconsciously at the wrong time by the people or things around us. Sometimes we're like the students walking slowly down the corridor because someone had unconsciously reminded them of old age; or we're like the students drinking too much beer because they want to socialize rather than because they really want more beer. It can all mean we end up performing behaviours that don't line up with our long-term goals.

The problem for making and breaking habits is that so much is happening in the unconscious mind.[7] Since the unconscious is generally like the Earth's core, impenetrable and unknowable, we can't access it directly. This means that deeply held goals and desires can come into play without our realizing. Not only this, but our conscious intentions to change can prove too weak in the face of the behaviours we perform efficiently and automatically, with only minimal awareness.

What does all this mean for our attempts to control ourselves and our chances of making changes? It's this question that dominates the rest of the book. In the third and final section, we will look at how we can use our conscious minds in concert with the environment to make the lasting changes that we so desire. Before that, we look at how habits play out in everyday contexts. Here, discovered over decades of research on how we work, socialize, use the Internet, and morc, we find further clues to moulding the unconscious to our will and making lasting changes to habits.

PART 2

EVERYDAY HABITS

5

THE DAILY GRIND

Luke Rhinehart is bored. Really, unbelievably bored. He's a psychiatrist with a wife and two children, a moderately successful career, and a nice home. But he's fed up with it. He has become bored with all the repetition in life, the endless sameness of the days, and the activities in those days. He feels like he's explored all the regular avenues of interest that life has to offer and all he sees is mundanity and tedium. He has tried to find solace in therapy, existentialism, and Zen, but nothing works. Instead, he feels mired in routine, locked into the numbing repetition of everyday life.

One night, he stumbles upon what seems like a solution to the boredom of existence: dice. Instead of following his normal, everyday routines, he will periodically write down a list of six options, including actions he wouldn't normally choose; then he will roll the die and obey it, whatever the result. The force of habit is subverted both by consciously making a list of options and by allowing the die to choose randomly between them. Soon, life for Rhinehart has become anything but boring. In everyday situations at home, at work, and while travelling, he experiences the force of habit pulling him towards his normal routines and the dice pulling him towards randomness, chaos, and new experience. What he calls 'dice therapy' helps him fight the patterns ingrained into his personality to such an extent that he becomes liberated from his own self and, indeed, the idea of selfhood.

The character of Luke Rhinehart is the fictional protagonist of *The Dice Man*, a cult classic by George Cockroft, himself a psychologist, published in 1971.[1] In the novel, Rhinehart starts off by urinating into plant pots, walking backwards, and instigating 'Habit Breaking Week', then goes completely off the rails, leaving his wife and family, committing sexual assaults, murder, and eventually founding a cult. While we might not agree with the choices Rhinehart gives himself in the novel and the path he ends up on, the idea that the randomness of the dice provides an escape from everyday routine is certainly appealing.

The question is: if new experiences are so exciting, then why aren't we all dice men and women? Why don't we give ourselves up to randomness to escape the confines of habit? One answer lies in Cockcroft's novel. Rhinehart finds that once his experiment is under way in earnest, many of those around him are appalled by his new, seemingly random behaviour. That's because not only do we take comfort in our own daily routines, we take comfort in the routines of others. This is the other side of habits. The more we are exposed to experiences in life, the more comfortable we become with them, and the associated positive feelings can increase. It's why research shows that students sitting in lecture theatres tend to sit in the same seats, or as near to them as they can.[2] Even across different rooms they sit with their friends in similar configurations. It's also why air passengeres feel more comfortable about flying the more they fly: routine boosts feelings of safety.

It's not just behavioural routines that we feel more comfortable with; it's also intellectual routines. Routines reduce the stress associated with much-practised mental processes because they can be performed easily and unconsciously. Think of experienced A&E doctors who are able to stay calm when confronted with a dying patient by using habits drummed into them over the years. Routines can also provide a safety net when we're under extreme emotional stress, such as after the death of a loved one or some other major upheaval. We can find solace in the reassuring routines of everyday life: the regular tick-tock of getting up, going about our business, and heading off to bed at the same time each day, just as if nothing had happened.

Despite the comfort and security that habits can provide, they generally don't get a lot of good press. The expression 'the daily grind' doesn't exactly summon up visions of skipping through sunlit meadows on a warm, hazy midsummer afternoon. Rather, it makes us feel like robots; robots who get up in the morning, go to work, come back, eat, turn on the TV, spend 30 minutes worrying about tomorrow before falling asleep, only to get up the next day (often without enough sleep) and repeat the whole routine again. Where's the fun in that?

People do experience emotions while performing habits, but usually not ones that are related to the habit itself. Our minds are off somewhere else, and the emotions we experience are related to wherever our minds have gone. Perhaps this helps explain why one study has found that people are only half as likely to feel pride about a habit compared with a non-habit.[3] This suggests we don't associate our habitual behaviours with our ideal selves. This is odd since so many habits are good ones: being punctual, washing our clothes, or remembering the good times. In the same study, the researchers found that participants' thought habits weren't as important as non-habits in reaching goals, and were relatively uninformative about themselves and others.

Why would we think that almost half our daily behaviours (or maybe more) say little or nothing about us as people? Actually, these are just the sort of results we'd expect, given what we already know about habits. It's a result of their basic elements: that they are unconsciously cued by situations, that we sometimes have little control over them, and that they are performed with little emotion related to the habit itself. The effect of these factors is to give us a reduced sense of control over our actions. To us, from the inside, we can become detached from our habits, as though we are being externally controlled. In the same study, the one that asked about habits and pride, participants were also asked how much ownership they felt towards their habits. The responses were once again underwhelming. People were less sure of why they performed habits than non-habits; also, they thought they were less likely than non-habits to be caused by either the situation they were in or the other people around them. So, overall, people didn't feel that much of a causal connection between themselves and their routine behaviours, which accounts for the lack of pride and their negative emotions.

All of this routine can make us feel like rats stuck in a maze – and it's no coincidence that that's how we know the basics of habit formation. Researchers such as the American psychologist Clark Hull had rats running around mazes trying to find pieces of cheese, and from this he witnessed how habits were formed at the most basic level. As the rats repeated the same behaviours in the same circumstances, they got to the cheese quicker – much like people wandering around underground rail systems looking for their destination. It's all about stimulus and response. The stimulus is the way out and the response is to explore the tunnels and escalators looking for that magic exit sign. Of course, there isn't just an association being built up between stimulus and response; our actions are more conscious and goal-directed than that. They focus on an outcome, or, in this case, a way out. As we repeat our behaviours, what we're learning is an association between our actions and their outcomes. For instance, if I take the Northern Line from Waterloo and get off at Goodge Street, I'll find my way to University College London. In animal learning terms, a habit is formed when we move from *action-outcome* links to *stimulus-response* links.[4] In other words, a habit emerges when we pay less attention to the outcome and simply respond to the environment – usually in the same manner in which we have done so before.

Daily habits can make us feel like a rat stuck in a maze because we are behaving in a very similar model to the rats. This might sound a bit demeaning to humanity, but remember we're talking about the automatic, unconscious aspects of our behaviour. After all, the everyday habits we learn do some great things for us. Without the parts of our brain that help make habits, our lives would be that much more difficult. People with Parkinson's, a degenerative brain disease, have exactly this problem. The disease causes a decrease in the neurotransmitter dopamine in a part of the brain called the basal ganglia. This structure is important in controlling our movement, hence the characteristic shakiness of people with Parkinson's. But the basal ganglia is also thought to be vital to how we form new habits. With impaired neurotransmitter function in this part of the brain, Parkinson's patients have difficulty learning new habits and can even forget old habits.

This deficit has been demonstrated in an experiment conducted by Barbara Knowlton at the University of California, Los Angeles and her fellow researchers.[5] They recruited some normal control participants, some participants with Parkinson's and some participants who had memory problems. All of the subjects carried out a task in which they had to try to predict the weather (rain or shine) from four mysterious cards adorned with various geometric shapes. Participants were presented with the four cards in different orders, over and over again, then told what sort of weather was associated with them. The task was set up to make it very difficult to work out consciously the association between which cards show up and the outcome. But there *was* a pattern which the control participants picked up unconsciously. After a while, they improved from 50% accuracy (pure chance) up to 70%. The people with Parkinson's, though, didn't improve because they couldn't learn the unconscious association, they couldn't learn the habit. The amnesiac patients, on the other hand, did just fine. The test didn't rely on being able to remember things consciously, only that the unconscious was in good working order. For Parkinson's patients, unfortunately, it wasn't. This is the kind of unconscious habit-learning that most of us rely on every day of our lives to help get us through the most mundane of situations, like using a parking meter, operating our mobile phones, or making small talk.

Social Habits

Let's do a little childhood reminiscing. Think back to the meals you had in a specific house at a specific table with your family. Can you remember where you sat? Can you remember where your mother and father and siblings sat? Many readers, like me, will see themselves sitting in a particular place with other members of their family always sitting in the same positions. I remember in one house my Dad always sat to my left, my Mum to my right, and my sisters opposite. The routine was different when we moved to another house; there, my Mum still sat at one end of the table, but my Dad sat opposite me because the head of the table was flush against the wall. If I think back to either of these houses and try to imagine myself sitting on the other side of the table, it feels wrong, even now.

Perhaps you can remember some other aspects of family meals. Was there generally talking while you ate, or was that reserved for afterwards? Did you eat certain foods on certain days? Was Sunday lunch special? What happened when guests were there? Psychologists have studied family routines such as bedtimes, chores, watching television, Christmas celebrations, and family reunions, but it's the importance of mealtimes that comes up again and again. Despite increasingly fractured lifestyles, many families still believe strongly that they should eat together. This appears to be no bad thing, as evidence suggests that these kinds of family rituals are healthy. Families that have established good, predictable routines tend to be happier, with both parents and children being better adjusted.[6] Family rituals provide a kind of safety blanket; they increase the family's sense of togetherness and help build its identity. As a child, this sense of distinct routines and identities was never more obvious to me than when I visited a friend's house where they frequently did things so differently I might as well have been on the moon.

The habits of both behaviour and thought we develop as children can live on long after we've flown the nest. Early social habits in particular can have a striking effect on the rest of our lives. For example, take one of the most natural social habits most of us develop at a young age: being polite to strangers. We may occasionally be rude to people we know well, like our partners, our friends, and our families, but when it comes to strangers, we are usually quite polite. Those of us who have the habit were socialized into it at a very young age. Our parents encouraged us to say please and thank you to strangers, and over the years we began to notice that when we projected warmth towards other people, they treated us better. This a self-fulfilling prophecy because our habit of expecting acceptance leads to warmer behaviour, which in turn leads to greater acceptance by others.[7]

Not everyone learns this habit, though. For whatever reason – but probably a mixture of genetics and circumstances – some learn at an early age to be very pessimistic about other people, so pessimistic that it doesn't seem worthwhile to think positively about strangers. Some people learn to expect rejection from others and so, ironically, that's exactly what they get. Whether we're socially pessimistic or optimistic is a habit of social thinking that has huge implications for our lives. Just this one good or bad habit can help lead us towards either a lonely life or one filled with friends.

In the normal course of things, though, even those who are socially pessimistic will have picked up friends through school, work, and other interests. But how do we choose who we're friends with? The standard psychological account is that much of it is down to similarity: consciously or otherwise, we choose people who have similar backgrounds, tastes, attitudes, and preferences. Certainly many studies conducted in psychological laboratories around the world back this up. If you sit two random people together, they are more likely to express a liking for each other when there is also similarity between their personalities, cultural background, attitudes, and even their physical appearance.[8] This has long been the orthodox view amongst social psychologists, but it obscures something vital about the importance of our behavioural habits.

Think about a friend of yours, along with the context in which you first met, and where you meet to socialize now. For many people, both will be stable situations. For example, the friend you met at school, you now meet up with once a month for a meal or in a bar. Or the friend you met at work you now socialize with at home. Still, intuitively we tend to think that where friends meet and what they do together isn't as important as what they say and how they connect psychologically. But is the place and the activity as incidental to friendship as we tend to think?

Evidence that points more in the direction of habits comes from a study of friendship which measured the attitudes of participants along with what activities they took part in.[9] People were asked whether they liked things like hang-gliding, chess, football, or reading, as well as what attitudes they had towards things like religion, politics, and economics. Then the researchers used statistical methods to look at the connections between friendship and both activities and attitudes. They wanted to see where friendship was forged. Was it more what they did together, or was it more how much they shared similar attitudes? The researchers found that friends tended to share the same preferred activities much more than the same attitudes. In fact, the attitudes of friends had no more in common than those of strangers.[10] This was the exact reverse of what people expected, which was that their attitudes would be more similar than their activities. As the subtitle of the study memorably encapsulated it: 'Those who play together, stay together'.

It becomes clear that what we do in similar circumstances is tightly bound up with our social ties. It's hard to tease apart whether mutual liking or mutually enjoyable activities come first, but we probably tend to underestimate just how important our shared habits are in forming and maintaining our friendships. No doubt our inability to notice just how important our habits are in our friendship ties is down to their unconscious nature. We frequently find ourselves in the same contexts, with the same people, talking about the same things, without specifically willing it and often without noticing. And that's not a bad thing. Without the tendency to associate with others in regular contexts, we'd miss out on one of life's great pleasures. In some ways, love itself is a binding together of habits. Two people's habits become intertwined so that, as interdependence increases, both people benefit from the routine activities within the relationship.[11] What could be a more admirable habit than love?

Work habits

Like almost every other area of human existence, the workplace is also a hotbed of habits. At work, habits do a lot more for us than we might imagine. It's not just about getting through the mundane parts of our working lives successfully, like commuting, making routine phone calls, writing routine emails, or making the right noises in meetings. According to one influential academic account, routines are at the heart of how the economy works because they store knowledge, provide stability, reduce uncertainty, and help people work together.[12]

Perhaps most importantly, routines allow people to coordinate with one another. When people work together towards a common goal, they need to have a rough idea of what other people are doing, how long it will take, and what they will produce. Without these kinds of routines, goods will arrive at warehouses with no one to unload them, company reports would have missing chapters, and both children and teachers would keep missing their classes. The fact that we (mostly) turn up on time, and the work we've done ourselves (mostly) meshes with others, means that all manner of social institutions can continue to function. Without routines, work would be a comedy of errors.

But working together isn't just about turning up on time; it's also about learning how we do things around here. When you start a new job, you can read all the manuals and know theoretically how things are supposed to work, yet nothing beats experience. The simple reason is that a lot of the rules we work by are unwritten. That's why many employees say their managers don't know the half of what goes on in their organization. Sometimes this is called organizational culture, but it's really about learning the small routines that make the organization work.

One fascinating example of how important habits are at work is in bicycle manufacture. Although we now think of the bicycle as a rather ancient and well-established technology, around the turn of the twentieth century it was far from that. Bicycles came in all sorts of designs: originally, they had no chains, tyres or gears, and the wheels were of different sizes. This made them difficult to ride and quite dangerous, as they tended to throw you over the handlebars (this was quaintly called 'a header'). As companies worked towards what became known as 'the safety bicycle', the industry went through all sorts of foments. With the prospect of big profits, the number of manufacturers exploded, as did the rate of innovation.

In their analysis of the bicycle industry between 1880 and 1918, Glen Dowell and Anand Swaminathan wanted to see which types of innovation worked best.[13] Was a manufacturer that completely reinvented the bicycle in huge leaps forward more likely to survive, or was the better strategy to go slow and 'get it right'? What they found was that manufacturers who tried to change too quickly were most likely to go out of business. The same was true of the manufacturers who were too slow to change. The ideal pace was in the middle. The bicycle manufacturers who were most likely to survive were those that kept hold of their old, good routines as well as trying to establish new ones. These companies continued to produce their old product lines at the same time as their new ones (in this case, the cut-off was at about four years – after that, companies began to stagnate). Workers could then move smoothly, over time, to the new production line, but they brought their old habits with them. Companies who overlapped their production could also keep their routine relationships with suppliers, distributors, and customers, which further helped them prosper.

So in efficient companies, routines and habits can and do evolve. Rather than being stuck in their rut, workers adapt to the new circumstances, slowly but surely, while bringing along the vestiges of their old habits. Research in a variety of industries such as medicine, technology, and car manufacturing bears this out: routine behaviours don't hold completely still in institutions and organizations.[14] As with all habits – organizational or personal – it all depends on how we deal with the feedback from the environment and whether we are ready and able to make changes.

Travel habits

The number of habits involved in the simple act of travelling is huge. Take driving: habits mean we automatically operate the indicators, turn on the radio, respond to the car in front braking, and anticipate the actions of other drivers – all on top of navigating to the supermarket. Without habits, the drain on our memory and our decision-making powers would be too great. Consciously having to recall how to brake and which road to take would take an enormous toll. We'd have a stroke before pulling out of the drive.

At the societal level, our driving habits create all sorts of environmental problems. For instance, the cost of petrol is astronomical, cars are expensive, and, in London especially, the roads are crammed with cars. Famously, the average speed of traffic in London is somewhere around 10 mph, perhaps slower, which inspired the marvellous headline: 'London cars move no faster than chickens' (although that is chickens *running at top speed*).[15] Of course, many journeys are necessary, but especially in cities, there are all sorts of cheaper and more efficient alternatives to cars. Given the gridlock, why do people continue to drive?

Whatever the political or environmental arguments, from a psychological perspective people's behaviour needs explaining. Travel choices are powerfully shaped by habits because they can override our intentions to change, but where do these habits come from in the first place? A clue comes from a German study which looked at what influence socialization had on whether people chose to use their cars for transport.[16] Almost 4,000 students at a German university were asked questions about how they travelled. Across two studies, the researchers looked at a variety of factors that might influence what method students took to travel to university. Amongst other things, they were asked about how much their parents used public transport, how much they thought a driving licence was an initiation into adulthood, and how open-minded their peers were about different forms of transport. The results showed that each of these factors was only important in travel choices insofar as it tended to create travelling habits. In other words, students were socialized into, say, car use by their parents, and this early build-up of habitual behaviour then took over. This shows how quickly it is possible to inherit travel habits, and how immune they soon become to things we think might influence them, like peer pressure and feeling in control of choices.

This isn't to say that all our travel choices are habitual; many are not. The problem with habitual travel choices, though, is that the more ingrained they become, the less likely they are to be consciously re-evaluated. For example, we might continue to drive to work in the summer, despite the fact that cycling is much cheaper and healthier, because of the habit we've built up during the winter. Or we might automatically plan a new trip by car, because we always use the car, without even considering the available alternatives, like trains or coaches. Lab research finds that if people habitually drive to shops that aren't within walking distance, they don't usually consider going on foot, even when it's a shop that is within walking distance.[17]

Encouraging people to change their travel choices is very hard. Research suggests that the key is getting people to reconsider their options consciously before they automatically get into the car.[18] Clearly, this is much easier said than done. One method that's been tested is handing out free bus passes, which can work to weaken driving habits.[19] Still, any method that's used needs to take into account the fact that while travel choices may originally have been made for rational reasons, once they become automatic and therefore unconscious, rationality goes out of the window. To break a habit such as travel choice, we need more than just the desire; we need a specific type of plan, which we'll come on to in the third section of this book.[20]

Eating habits

What could be more routine than eating? And, as everyone who is on a diet knows, all that eating adds up. If a man consumes 3,000 calories per day, then that's more than 1 million each year and getting on for a hundred million in a lifetime. Every day, we make all sorts of decisions about eating, like where, when, what, and how much. At work, we might decide to go to a salad bar rather than a fast-food restaurant, or we might decide to skip lunch altogether. At home in the evening, we choose between doing some cooking or putting a ready meal in the oven. However, because many of our habits are unconscious, we often don't notice the decisions we're making. Like our other daily routines, the effects of all these decisions accumulate over time and are there for all to see around the waist.

So how many decisions about food do we make each day, and how many are we aware of? Brian Wansink and Jeffery Sobal asked 154 university students to estimate how many food decisions they made every day.[21] The average turned out to be 14.4, which sounds like a reasonable number. But then the participants were pushed to think harder about it. They broke down their eating into five different aspects that any journalist will recognize: the who, what, where, when, and how (much). For example, when do you start and stop eating? Who's with you? Where do you eat? And so on. From this, they were able to make a much better estimate of the total number of decisions about food that people make each day. This average was a mind-boggling 226.7, which is frighteningly high in comparison to their initial estimate. To check this, three participants were given a clicker which they clicked every time they made a food-based decision over a 24-hour period. Their clicks suggested that the number of daily decisions about food came to over 200. This study is a good demonstration of how our eating habits, like our work, socializing and driving habits, fly under the conscious radar.

Now a critic might say that there's a bit of cheating going on here, to inflate the difference between our estimated and actual number of food-related decisions each day. For example, people don't tend to think of deciding not to eat as a food-related decision, although, of course, it is. But even when only decisions to eat food positively were counted, the average was still 59 decisions per day. And as you'd expect, obese people made more food-related decisions, presumably because they were eating more food. Wansink and Sobal call this 'mindless eating', and Brian Wansink has written a fascinating book with that title.[22]

What this means is that many of our eating habits take their cue from quite simple habitual decisions we make, often without our realizing. You can see how little we know about what we eat in some circumstances from a study of popcorn-eating in a cinema. This found that people given a 240-gram box of popcorn ate 53% more popcorn than those given a 120-gram box.[23] It isn't that surprising that when given more food, people ate more. What is surprising is that when asked afterwards to estimate how much they had eaten, only 6% thought that they'd eaten more than usual. Then, when told about the larger container, only 5% of people thought that it had influenced their eating, while 77% said they were hungry and 15% flatly denied they ate more.

In another study, the researchers found that, even when they made the popcorn less palatable, it had less influence on consumers than the size of the container they were eating from.[24] We tend to think that how much we eat is mostly affected by the food's quality and our appetite; these studies suggest otherwise. In fact, we are taking relatively trivial cues from the environment, and, along with our habits, these are having a large impact on our intake. The examples go on and on – the size of plates, the people we are with, whether we're watching television, and so on – but the general point is the same. The situation we are in cues our eating habits, and we don't notice the choices we're making because they are unconscious.

Shopping habits

In the past few decades, a revolution has happened in how companies market their products to us. This revolution is partly a result of advances in computing and the ability to capture and process data, and partly of a key psychological insight. By now, this revolution is far from secret, as most of us carry around the evidence in our pockets: loyalty cards.

Before the advent of loyalty cards, companies marketed their products to us on the back of some limited research based on putting us into one of a number of broad categories. For example, one market segmentation system called VALS (Values, Attitudes, and Lifestyles) splits consumers into groups: innovators, thinkers, believers, achievers, and survivors, amongst others.[25] In theory, each group is thought to have a different mix of motivations and resources. In reality, of course, the categories are crude and can't hope to represent everyone in all of their diversity.

Then, along came store loyalty cards and the ability to collect and process huge amounts of data about people's shopping habits. Soon enough, some retailers, like the supermarket Tesco, were discovering that this data was a rich mine of information. They could work out in which stores to launch new, expensive ranges of foods and which areas of the store particular customers didn't habitually use. For large companies, the identification and leveraging of quite subtle habits and patterns in the data could make them billions of dollars.[26] Where we were once in vague and often illusory marketing categories, as consumers we can now be brought into sharp focus with a glance at our purchase history. With the data that online retailers are now collecting on us, this is only the beginning.

Part of the reason loyalty cards can be so useful for marketers is that they capture people's shopping habits, and our habits are surprisingly predictive of our future purchasing, sometimes more so than our own intentions. In one study that tested how shopping habits stack up against intentions, participants were asked how often they bought food from a fast-food restaurant.[27] They were then asked about their intentions over the next week: how often did they intend to buy fast food? What they found was that when people's habits were weak, their intentions strongly predicted their behaviour: in other words, without an established habit, people bought what they intended to. However, when the habit was strong, intentions were only a weak predictor of behaviour. So, in the face of a strong habit, we sometimes don't buy the things we intend to; instead, we just do what we did before.

The idea that strong habits rule our intentions goes against our intuitive sense of how shopping works. It feels to us as if we choose a product because it provides the best trade-off between what we're prepared to pay and what it offers. Since we're satisfied with our initial purchase, we go and buy the same thing until we're no longer satisfied. Then, perhaps the quality drops or there's another cheaper product, and so we change our choice.

This story we tell ourselves is the exact same story the marketing people tell each other. If you read a marketing textbook, they are filled with page after page on customer loyalty and satisfaction. Satisfied customers are loyal customers, and loyal customers are profitable customers. People are assumed to learn from their experiences. Poor evaluations are thought to lead to lower satisfaction, which leads to a change in purchasing habits.

Although this is such a familiar story for marketers and

consumers alike, there are all kinds of hints that this isn't the way our mind really works. If satisfaction was the key to keeping customers loyal, then why are between 65% and 85% of people who switch brands either satisfied or very satisfied?[28] The fact is that marketing professionals have very little idea what makes people switch from one brand to another.[29] On its own, satisfaction predicts very little of people's behaviour, perhaps as little as one-quarter.[30] For the student of habits, this makes sense since it's easy for habits to become divorced from intentions. When the initial choice is made to buy a product and subsequently made again and again, in the same context, then it's likely to become a habit, meaning we make the same choice without considering the options. For many purchases, this isn't a problem, of course, and saves us time. Where brand loyalty does become important is when purchases are made in new and different situations. For example, choosing a new car is unlikely to be a habitual purchase.

All this leaves us with a rather disturbing answer to the seemingly simple question of why we buy what we buy: much of what we buy regularly, we buy because we bought it before. While these products might once have met our requirements, and we might once have thought carefully about them, there's no reason they still do – our needs, and what's on offer, may well have changed without our noticing. Worse, we are likely to cover up this fact by automatically justifying our own decisions to ourselves, and because we often have little access to our own unconscious, it's easy to become largely 'unconscious consumers'.[31]

The very fact that habits are so easy to perform makes them attractive. People are what psychologists call 'cognitive misers': for the most part, we prefer to avoid difficult decisions. So it's partly the mental costs of switching that put us off. Habits, on the other hand, can literally feel good. When we can quickly choose a product with minimal thought, we feel more successful. Each time we buy it and use it, we become more comfortable with it and the other options fall further and further behind (unless the product or service is truly awful).[32] Set against this, we avoid casting the net wider for other options,[33] and deliberation hurts the brain: not much, but just enough to make us wonder whether a slightly better-tasting peanut butter is really worth the effort.

Habits can be so strong that they don't respond to the rewards marketing people throw at us to encourage switching. Even the standard economic idea that consumers are rational and will respond to rewards is suspect. That's why research finds that habitual shoppers often don't respond very strongly to incentives like special offers.[34] Habitual consumers are so locked into their habits that they don't have much effect.

Of course, not all of our shopping is habitual. Some people are happy to switch their patronage to wherever the deal is best after putting in the necessary effort. Most of us, though, probably have a sliding scale, with some of our shopping being habitual and other parts not. While we're more likely habitually to buy the same brand of milk from the same shop each week, buying a new phone is unlikely to be a habitual purchase.

What if you're staring into the kitchen cupboards and wondering why the same old products are always staring back at you? Or what if the monthly grocery bill has one too many zeros at the end? There are three factors which are most likely to make us robotic, habitual consumers: being short of time, being distracted, and having limited self-control.[35] It's a double-edged sword: relying on shopping habits saves us time and effort and lets us think about something else, but we also get the same things we always get. Ordinarily, this may not be a bad thing, eating the same things all the time, because you're tired and distracted and shopping is, well, a recipe for boredom. The trouble is that old habits are hard to unlearn.

Sometimes, though, we do get shaken out of our old habits by changes in our lives. This was first demonstrated by the consumer-behaviour expert Alan Andreasen, who discovered that there seem to be particular moments in people's lives when their consumer habits are most ready to change.[36] He interviewed hundreds of people and asked them whether they had changed their usual brands in the last six months. He also asked people whether they had experienced any major life events in the last six months, either positive or negative. When he looked at the data, what he found was that the more major life events they'd experienced, like changing employer, getting married, moving house or school, the more they had changed brands. Not only that, but people who had changed were also more likely to be satisfied with the switch they'd made. Subsequent research has also found that major life changes are likely to lead to increased variation in consumer habits.[37]

The academic explanations for these changes are like the stories we tell ourselves: often quite complicated and sometimes misguided. The truth is probably simpler. Major life changes mean changes in situations, which means old habits are disrupted. After moving house, you visit a different supermarket with a different layout and suddenly, choices that have been largely unconscious, cued by the situation, are offered up for deliberation to the conscious mind once more. Or, after switching employers, you earn more money and decide to move up to a more expensive brand. In either case, willingly or otherwise, your conscious mind is back in the hot seat making all those crucial decisions about which brand of coffee to buy or where to take your dry-cleaning.

~~~

Most of us, at some point, find the daily grind is getting us down. It's not just work, but our routine feels old. Sometimes we want to break out and do something different; we feel like seeing some new sights, meeting some new friends – somehow changing the old routines. But anyone who has tried this knows that ingrained habits are hard to budge, and they get harder as we age. Habit provides a safety zone, but it's also a kind of cage from which escape is hard.

One reason habits are so hard to change is that we start performing them without conscious deliberation. When we shop, socialize, eat, work, and travel, it's often difficult to spot that we're responding in the same way to the same situation, and sometimes, that that response could be improved.
~~~

The thread running through all of these everyday activities is that the first step is simply to notice our own habitual behaviour. For organizations to change and improve their habits, they have to notice and respond to what they are doing. For people who want to change their eating, the first step is to notice what they are already eating. For those who want to change how they travel, the first step is to stop and think about the choices rather than follow the same old routine.

Everyday life doesn't have to be such a grind, if only we could notice our own behaviour. Perhaps surrendering ourselves to the randomness of dice is too extreme a solution, but certainly there are ways of mixing things up that would improve our everyday existence. Like the bicycle manufacturers who successfully survived the industry cull, we need to try to spot which routines are helping us and which routines are killing us, and then try to work out how to replace them. But slowly.

6

STUCK IN A DEPRESSING LOOP

Around the age of eleven, Stanley became obsessed with symmetry. Suddenly, everything he did had to be symmetrical. His writing had to be perfect, with all the lines, loops, and dots just so. When he walked down the street, his arms had to swing exactly in time with the stride of his legs so that he seemed to walk like a robot. When he took tests at school, he became more obsessed with shading the answer boxes precisely within the lines than with getting to the end of the test. Later, certain numbers started to go round and round in his head, particularly the number 6. He began to repeat everything he did at least twice: he went back over his paper round again and again to check he hadn't missed a delivery, and soon these habits were taking over his life. Stanley called these obsessions 'mosquitoes of the mind'.

Stanley is one of the sufferers of obsessive-compulsive disorder, OCD, documented in Dr Judith L. Rapoport's book *The Boy Who Couldn't Stop Washing*, and he is far from alone.[1] The number of people who have it at any one time may be as high as 2%.[2] This means there could be over 1 million people with OCD in the UK. OCD is classified as a kind of anxiety disorder; people who suffer from OCD have obsessive or habitual thoughts about, say, being clean, which drive their repetitive behaviours, like hand-washing. People with OCD are almost always aware that they are obsessive and that they are repeating the same behaviours over and over again, yet they find it difficult to stop. OCD is a disorder of intrusive, repetitive, anxiety-provoking thoughts. The compulsive behaviour is carried out because it helps to relieve the anxiety associated with the thoughts.

People with OCD have all sorts of obsessions and compulsions, but across cultures and countries three are practically universal: cleanliness, checking things are done, and imposing order. These are not exactly wild and wacky obsessions that are difficult to empathize with. Who hasn't wanted to tidy up the house, wash their hands, and check that they've turned off the oven? Perhaps you've even cleaned, checked, or washed a little more assiduously than might have been called for? Research tells us that most people experience unwanted thoughts, images, or ideas flashing through their heads from time to time, and some more than others.[3] This makes sense since there are all sorts of excellent reasons why it's handy to be clean, well organized, and on time. Even people who are always late, untidy, and disorganized know it's not ideal!

What is it that makes regular, everyday worries cross a line and become a disorder? We don't know exactly, but there are convincing theories. The problem for people with OCD is that relatively normal fears, ideas, or images aren't ignored or discarded as random thoughts that the mind often creates. Instead, a worrying image, like stabbing someone with a knife, which may start only as a brief flash of an idea crossing the mind, takes on much more importance than it should. A person with OCD can become fixated on this type of image. They can come to believe that they might act on it. To try to counteract this perceived danger, the person might compulsively avoid knives and continually check on the safety of their child. This is a perfectly rational response to an irrational thought, and the response also works, after a fashion: the worrying thought is dispelled, although only for a time. The obsessive habitual behaviour works like a safety blanket, comforting the sufferer. But the very fact that the compulsion relieves the anxiety about the obsession sets up a reinforcing loop. As the habit grows in strength, the mental state of the sufferer spirals downwards.

Not only do OCD sufferers have to cope with their habitual obsessions and compulsions, they also frequently suffer other mental health problems. Two-thirds will also experience depression at some point in their lives, and up to 90% will suffer from one other major disorder.[4] This underlines a really important point about mental health problems:

they often don't fit neatly into diagnostic boxes. People who are depressed are also frequently anxious, people with obsessive-compulsive disorder are also often depressed, and so on. As a result, the modern view is that these disorders exist on a continuum.[5] But the overlaps don't just occur between different types of mental health problems; they also overlap with (so-called) normal people. It's not just people with full-blown OCD who can get thoughts stuck in their heads.

Quite naturally, therefore, the language of psychology, psychiatry, and especially psychoanalysis has permeated everyday life to a huge extent. Many of us are now familiar with phrases like 'acting out', 'venting', 'bipolar', and 'borderline'. While perhaps not everyday language, they've gained much more recognition in recent years, so much so that the terms are now used in technically inaccurate senses. People may say things like, 'I get totally OCD about cleaning my kitchen'. When I conducted a poll on my website, this phrase was voted the most irritating example of psychobabble.[6] I have some sympathy with those who say that this sort of usage is offensive to people who really are suffering. But in another way, this type of linguistic slippage between how we talk about 'normal' and 'abnormal' mental functioning is telling. It hints at how narrow the gap is between one and the other, and it indicates something about all of our own everyday experience. The obsessions of people with OCD may seem weird and extreme, especially to those who experience them, but they are rooted in something fundamental about what it means to be human. We all of us worry and we all want to feel safe. Because habits can provide this reassurance – but can also get out of control – they are both savior and curse.

~~~~~~~~

People with Tourette's syndrome experience compulsions that are somewhat similar to those in OCD. Tourette's, which is most common in childhood, is often associated with involuntary swearing, but only about 10% of sufferers actually exhibit this. The similarities between OCD and Tourette's aren't surprising, since the two disorders have genetic similarities and are often co-morbid: people who have Tourette's also often have OCD.[7] Sufferers report feeling an urge building up inside them which has to be dissipated by a behaviour, such as an outwards stabbing of the elbow or some other movement of the body. The tics can be set off both by internal processes and by aspects of the environment, and the ticcing gets worse under stress. Like OCD, Tourette's sufferers find it very difficult to stop these unwanted actions. And like OCD, Tourette's has been linked to the basal ganglia, a structure buried deep in the centre of the brain that is important in habit-learning. Because it has a strong biological component, drugs are often used to help with the symptoms.

The tic of someone with Tourette's is like an extreme habit. Some of the most common tics are also very common social signals, like raising the eyebrows, nodding the head, or throwing up your arms in exasperation. Indeed, the tics may well be the result of what happens when social habits
~~~~~~~~

get out of control. Like regular habits, tics are often triggered unconsciously, may also be responses learned to particular situations, and, like regular bad habits, sufferers are desperate to change them. The difference is that for those with Tourette's and OCD, the tics and habits they perform are often much more difficult to inhibit and are much more inconvenient in everyday life.

Although serious sufferers of Tourette's and OCD often take medication, psychology can also help. One psychological intervention that is often used is habit-reversal training. Although it's designed for people with Tourette's, there's much to learn from the procedure that therapists go through for anyone trying to change their habits.

The first stage of habit-reversal training is all about awareness of the tics themselves. People with Tourette's are asked to think about their tics, often with the help of a video of themselves. Then they try to work out what internal or external factors might set off the tic. It could be something happening in the environment, like playing computer games, or it could be an internal thought or feeling, like thinking about Batman. The idea is that with greater awareness, there's more chance of being able to catch the tic building early in the sequence. In just the same way, thinking about and trying to notice our habits – both good and bad – is a great exercise for everybody. Even more crucial, though, is spotting the circumstances in which we perform our habits. Without knowing when they happen, it's difficult to make a change.

Stage two is called 'competing-response training'. This is

where tics are replaced with something else; so a 'bad' habit is replaced with a less bad habit. For example, many people with Tourette's have the tic of jerking their head to one side. So instead, they learn the competing response of tensing their neck muscles. Alternatively, people whose tic is sniffing will be encouraged to breathe deeply through their mouth. You might well ask why therapists don't try to replace it with nothing. Well, research suggests this doesn't work as well, and later on, we'll discover a good reason why. The practical upshot of this therapy for changing habits is that we need to develop a competing response. Bad habits may be hard to change, but they can be sidestepped. That's why smokers chew gum: it's difficult to smoke and chew at the same time.

Those two steps are at the heart of the therapy: first, notice the habit; then introduce a competing response. While these can start the change happening, they can't necessarily keep it going; for that, you need motivation, willpower, and the ability to stick at it. This is hard for all sorts of reasons, but this therapy gives us two vital starting points: noticing a habit to be changed and replacing it with another behaviour.

Despite the difficulties in changing extreme habits, there's evidence that habit-reversal therapy can work for Tourette's sufferers. A randomized, controlled trial lead by John Piacentini from the University of California at Los Angeles recruited 126 children with moderate or severe tic problems and gave half the habit-reversal therapy, and the other half the regular supportive information and education.[8] Just

18.5% of the control group were either improved or much improved after 10 weeks, while in the habit-reversal therapy group, 52.5% were either improved or much improved. This shows that with a concerted effort even extreme ingrained habits can be changed. But note that this is no quick or easy fix: over the length of the study, children had a total of 10 hours with therapists along with help and support from their parents (plus, one-third were also taking strong anti-psychotic drugs). Fortunately, most of us aren't in an all-out war against our own biology; we're just battling old routines. Surely, if these children can successfully fight their disabling tics, we can manage much easier shifts in our lifestyles?

~~~

When we feel blue, the whole world looks blue, however it appeared yesterday. Bad moods colour our perceptions, changing our experience and plunging us into pessimism and darkness. The next morning, though, for most of us the world looks a better place. Where yesterday we saw only decay, ineptitude, and loss, today we see opportunity, success, and hope for the future. For some, though, the new dawn rarely comes. Every day, the world is still coloured blue, and there seems little point in carrying on.

There's a very simple question at the heart of depression: If bad things are happening to people all the time, why do only some people become depressed? In fact, some people seem to cope with all manner of bad things happening to them and they quickly recover, while others go under at the
~~~

slightest hiccup. The rest of us reside nearer the middle and take our slings and arrows as best we can.

Like many simple questions, the answer is complicated and not fully mapped out. But clearly it has a lot to do with the way we think. Depression isn't just (if at all) a biological disease of the brain; it is a way of thinking about what has happened to us and why. The processes that are at the heart of depression are, in many ways, a series of habitual thoughts. Here, we'll look at two that are most central to depression.

The first problematic habit of thought in depression is to do with appraisal. Appraisal refers to the way in which we try to find meaning in the world. Human minds are 'meaning machines' – we're always trying to work out why things are happening to us. The way we routinely think about why things happen has a profound effect on our view of the world and how we experience it. Let's say, just as an example, that you lose your job tomorrow. After updating your CV, you start looking for a new job. But that proves difficult, so you are unemployed for a time. This is the kind of thing that is likely to make people depressed, and your habitual ways of thinking about this situation in part determine whether you'll actually get depressed.

Let's say that you interpret losing your job in the following way. You think, on balance, that you lost your job because the economy is weak at the moment and your organization has been forced to cut staff. Although you've been unemployed for a while, there's a good chance that if you keep applying, you'll get something before long. You won't be surprised to learn that this is a very non-depressive way of responding. The meanings you've attributed to the event are as follows:

> *It's not my fault I've lost my job; it's because of the economy. In other words, the cause is external: I didn't lose my job because I'm no good at it.*
> *The situation is temporary: I think the economy will recover.*
> *I have control over the situation: I think that if I keep trying, something will turn up.*

Now, these views of the world may or may not be objectively realistic, but what we're interested in is subjective reality, what it feels like to be you. For a person with a negative attributional style – a pessimistic habit of thought – the thinking would be the other way round:

> *It's my fault I lost my job.*
> *I will never get another one.*
> *I can't do anything about it.*

It's pretty clear why thinking like this might make you feel hopeless. When you habitually think about negative events as though the causes are your fault, but also out of your control and will continue for ever, then it's likely to lead to a depressive state of mind. Studies continue to support this idea that depression is at least partly a result of this habitual way of thinking. One large study assessed the thinking styles of 5,000 students and followed them up over two years.[9] From this sample, 173 students demonstrated this particularly pessimistic way of making sense of the world. Of these, 17% went on to have serious depressive episodes. In the remainder of the students, the rate was only 1%.

If this attribution bias can help explain why some people suffer from depression, how can we explain why the rest of us don't? Life can certainly be depressing, but many of us only succumb to low moods for a while and then we recover. In other words, what habits of thought keep the rest of us mostly out of the woods? One very strong candidate is a mirror image to the negative attributional style, which is called the self-serving bias. We know that most of us use this habitual way of thinking because psychologists have seen the effects in decades of research. In these studies, people are asked things like:

> *How charitable are you compared to other people?*
> *How kind are you compared to other people?*
> *How lazy are you compared to other people?*

What we find is that people repeatedly rate themselves somewhat more highly than others around them. We tend to believe we are more charitable, cooperative, and kinder while being less lazy, deceitful, or belligerent than other people. People display the self-serving bias in both their abstract traits as well as in their specific behaviour. If you get people to predict whether they would help a stranger who had fallen in the street, they will tell you they are more likely to help than others. This can't be true, of course; we can't all be above average. Many of us are only average and almost half of us, by definition, are below average.

One piece of research has added together the results of 266 of these types of studies on the self-serving bias from around the world.[10] They found that the overall impact of the effect is large, meaning that it's very easy to spot. The self-serving bias is also stronger in people from the Western world, and particularly from the US, than it is amongst Asian people. Nevertheless, overall, the habit of thinking positively about yourself in relation to other people is strong wherever you look. Except in one place. People who are depressed don't tend to have a self-serving bias – in fact, they see themselves relatively accurately, which may well be part of the problem. That's not because depressed people are bad people, far from it; just that the rest of us seem to need to see ourselves with rose-tinted glasses in order to get by.

Another habitual way of thinking that seems to be important in depression is called rumination. Imagine I put you in a room, turn on some depressing music, say some Radiohead or Barber's *Adagio for Strings*, and then I get you to read a really depressing story. The story describes the unexpected and painful death of a loved one (this is one hell of a party). Afterwards, I get you to mull over the thoughts that are going through your head, really think about how you're feeling, the loss of a loved one and what it would mean (are you having fun yet?). This is exactly what researchers did in one early study.[11] The ruminating group were compared with another group who did a distracting task afterwards designed to get their minds off the depressing music and story. You'll be unsurprised to learn that the people who were distracted were much more likely to recover quickly from the depressing story and the depressing music. This study was carried out on people who were not clinically depressed, but this habitual way of thinking is often seen in people who are depressed and anxious.

In some ways, rumination looks like a good strategy for dealing with depressing things that have happened to you. People who use it frequently say they are trying to understand and solve their problems. The research shows that, in reality, rumination can interfere with problem-solving rather than help it.[12] This is partly because, as we've seen before, we have little access to our own unconscious, along with many of the real reasons why we think and behave the way we do. Of course, ruminating and distracting yourself are not the only habitual ways of dealing with difficult circumstances in life. You could try and suppress the thoughts, or you could do your best to avoid them. One study has compared rumination with these strategies and others across 114 different studies and thousands of participants.[13] They found that it's rumination that is most strongly associated with psychopathologies in general, including eating disorders, addiction, anxiety, and depression. However, while habitual ways of thinking are central to depression, anxiety, and other mental health problems, they can't explain everything. What they do is provide excellent examples of how habitual ways of thinking can either bog us down or, in the case of the self-serving bias, help pull us out of trouble.

One of the most effective modern non-pharmacological treatments for depression and anxiety is cognitive behavioural therapy (CBT). At the heart of this type of therapy is the idea that both depression and anxiety can be alleviated by changing habits of thought. The therapy assumes that certain situations can repeatedly and automatically cue unhelpful negative thoughts. Here are some examples:

- *Black-and-white thinking*: believing that if you don't achieve perfection at something, you're a failure.
- *Personalization*: assuming that when bad things happen it's all your fault, rather than just being bad luck or one of those things.
- *Catastrophising*: jumping straight to the worst-case scenario from relatively limited evidence.

There are many more habitual thoughts that are common. Most of us think these things from time-to-time, but for people who are depressed these thoughts have become overbearing and have started to take over their lives. When you jump to the worst possible conclusion with only the slightest provocation, life can be extremely depressing.

Like cognitive therapy for people with Tourette's, CBT first asks people to try to identify these automatic negative thoughts. As we've seen, this is difficult because they can be automatic and unconscious, but with a trained therapist, it can be done. Then, CBT uses a process of questioning to see if the habitual thought is a reasonable one. People are asked to think back over their past experience and see whether the disastrous consequences they envisage have happened before. Note that the goal of CBT is absolutely *not* about trying to 'think positive' or put a thought out of your mind. Avoiding a thought is very difficult: what CBT does is to try to replace an unhelpful thought with something more helpful. It's trying to use the weight of the thought against itself, like a kind of mind judo.

The same process is used to challenge core beliefs about the self. We saw that depressed people habitually attribute negative events to themselves, believe that they are out of their control, and think they'll never change. CBT also targets these core beliefs with similar tools. A therapist will try to examine how they came to such a negative view of the self, to try to teach them to be more sympathetic towards themselves and to feel some hope that change is possible. The problem is that these habitual ways of thinking have been ingrained over years and are very hard to change; they are also much closer to people's core personalities, and so more painful to think about. Naturally, then, it can take a long time for someone who is convinced he is a failure to change his mind, even a little.

For those of us who would like to change more modest

patterns of thought, though, clearly there is hope. But even for less persistent habitual thoughts, the same challenges are likely to present themselves. Namely, (1) we are likely to be unaware of either some or all of our habitual thought processes; and (2) the habit will resist change because it is automatically cued by situations or other thoughts. The reason why we might want to change patterns of thought is simply that they aren't doing us any good. Just as it's easy for behavioural habits to hang around long after their purpose is forgotten, it's the same for ingrained ways of thinking. Many people have thought patterns that are long past their sell-by date, but which are stuck to out of pure routine.

Take self-esteem as an example. People with low self-esteem have got into the habit of thinking about themselves as worth less than other people. This habit is likely to stem from childhood and has also most likely been reinforced by others over the years. It may also be a self-fulfilling prophecy: a person acts as though he is worth less, and so is accorded lower status, which then reinforces his low self-worth. Breaking out of this kind of loop is difficult, but not impossible. It involves asking questions about why you think in a certain way, and then trying to test these thoughts against reality. Are you really as worthless as you think? Certainly not. And it is also about addressing unhelpful core attributions such as thinking that mistakes and failures are all your fault, will continue for ever, and can't be changed.

Not all habitual negative thoughts are bad and should be changed. Since psychologists are often faced with people suffering the depressing consequences of habitual negative thoughts, we know more about them than negative habitual thoughts that have neutral or even positive consequences. This may sound like a contradiction but it's not, because negative habitual thoughts can have positive consequences. In fact, the dividing line between positive and negative outcomes can be thin.

The simplest example is worrying. We've talked about worry in the guise of rumination and anxiety and their negative consequences, but worry can also be good for you. For example, studies have found that people who worry more perform better at work,[14] are better at dealing with stressful events,[15] and can do better at school.[16] They may even be more healthy as a result of being more likely to engage in health-promoting behaviours.[17] The reason for this is obvious when you think about it. People who worry about their performance at work are more likely to try and improve it, and the same goes for students' academic performance and health behaviours. So the question becomes about the *type* of worrying. Edward R. Watkins from the University of Exeter has suggested that one of the vital differences between constructive and unconstructive worrying is hidden in the character of thought.[18]

Let's imagine two ways of worrying. In the first, let's say you're worrying about your health. You've noticed some pain in your leg and so you start to worry about it. This prompts you to think about other bodily problems and the general problem of the body's weakness. You wonder abstractly how long you have to live and then, increasingly morbidly, how many people will come to your funeral. Meanwhile, the leg continues to hurt. Here's a second way: you start worrying about your leg, which prompts you to wonder if you've pulled a muscle while playing tennis. So you visit the doctor and get it looked at.

This is a crude way to make the point, but notice that in the first example the worry is abstract, whereas in the second it's a concrete sort of problem-solving worry. Also notice where the abstract worrying ends up: assuming that a pain in the leg is the beginning of the end. This shows how small differences in the way habitual automatic thoughts are directed can strongly affect the outcome. Things tend to go better if worry prompts problem-solving than if it prompts an existential crisis.

For some people, though, life is one big existential crisis. The stereotype has been well mined in comedy with two of the best examples being characters created by Woody Allen and Larry David. These are people who manage to make their persistent negative thoughts work for them. So-called 'defensive pessimists' put a lot of work into predicting how things will go wrong. They both set very low expectations about what will happen, and spend a lot of time thinking about the exact circumstances of their impending downfall. Defensive pessimists are like super-worriers.

Preparing for failure is not exactly a well-known strategy for success, but in the specific case of the defensive pessimist, it seems to work quite well. Like persistent worriers, defensive pessimists are motivated to problem-solve by their intuition that everything will go wrong. For people who are very pessimistic by nature, this strategy has been shown to work.[19] Note, though, that the studies show that it doesn't work for people whose habitual thinking style is more neutral or optimistic. Still, this work does suggest that even super-worriers – people with long-term persistent, habitual negative thoughts – can benefit from a problem-solving habit of thought rather than focusing on abstract worry.

~~~

When people think about bad habits, the first things that come to mind are potentially self-destructive behaviours such as smoking, drinking, or gambling. Although habits of thought are not always as obvious for all to see, they can nevertheless have a huge effect on how we experience life. What we've seen in this chapter is that people who have OCD, Tourette's syndrome, depression, and anxiety aren't as different from the 'normal' population as we are often led to believe. The negative thoughts at the heart of these disorders are also regular parts of everyday life for most of us, although to a lesser extent. Negative thoughts aren't necessarily bad in themselves – indeed, they're perfectly natural – it's all about whether or not they help us solve real-world problems or become useless obsessions that take over our mental lives. The fact that many habitual negative thoughts can have positive outcomes only reinforces how useful our habitual negative thoughts can be (think of defensive pessimists).

The treatments that have been developed suggest ways
~~~

we can challenge negative thoughts that have become problematic for us. These focus first on identifying the problematic thoughts (no mean feat in itself), and then on trying to adjust them. Cognitive therapies attempt to create this switch by re-examining the habitual thought in the light of our experience. We can ask ourselves whether our wilder fears are well founded, whether our aims are realistic, and whether our ways of interacting with the world could be improved.[20]

Of course, not all repetitive thoughts are negative. At the other end of the spectrum are positive habitual thoughts. We've looked at the self-serving bias as one manifestation of repetitive positive thinking; this is a habit of thought that helps us to see ourselves and those we love in a more positive light. There are many other types of positive thought; for example, repeatedly thinking how lucky you are, letting your mind wander back to happy memories, and having pleasant daydreams. With practice, we can train ourselves to perform these 'happy habits' regularly so as to break out of a depressing loop and improve our everyday experience.

7

WHEN BAD HABITS KILL

On 31 August 1988, Delta Air Lines Flight 1141, a regularly scheduled domestic flight, taxied for take-off at Dallas-Fort Worth International Airport. In the cockpit, it was business as usual as the last preflight checks were completed. Shortly before 9am, the Second Officer called out a standard part of the checklist procedure: 'Flaps', to which the First Officer replied, 'Fifteen, fifteen, green light.' These innocent-sounding five words exchanged between First and Second Officer are a vital part of the take-off procedures. The wings of a large commercial passenger jet are designed for optimum efficiency when cruising at 40,000 feet, but during takeoff they don't provide as much lift. To help launch a 70-ton jetliner into the sky, the wings have to be larger: this is done by extending slats from the front of the wing and flaps from the back. In the Boeing 727 they were operating, the co-pilot does this by moving levers in the cockpit and checking that a green light comes on to confirm they are in position. Without flaps and slats correctly set, the plane may be drastically short of lift when the pilot pulls back on the control column.

To observers on the ground, the take-off didn't look right. Within seconds, the nose was pointing too high and there were sparks coming from the rear of the plane. In the cockpit, the stick began to shake violently. This tells the pilot the plane is stalling: it's about to fall out of the sky. Ground observers saw the jet rolling from side to side, and on the cockpit voice recorder the captain is heard to say, 'We're not going to make it.' The climb soon faltered and, twenty seconds after take-off, the right wing hit an antenna, spinning the plane and sending it crashing back to the ground. In the impact and subsequent fire, of the 108 people on board, 14 were killed and 26 seriously injured.

When the accident was investigated by the US National Transportation Safety Board, two major causes were identified.[1] The flaps and slats had *not* been set to 15; in fact, they were at 0, which is for normal flight, rather than for take-off and landing. Also, there was a warning horn that should have sounded when the pilots tried to take off with the wings incorrectly configured. Unfortunately, this did not sound because it was broken. The reason they couldn't take off was simple enough: when the pilot pulled back, they didn't have enough lift. The question was how such a regular pre-flight checklist, which had been successfully completed so many times, could have gone so tragically wrong.

Many aspects of flying passenger jets are routine. Pilots have to get into the habit of performing all the necessary checks to ensure that their plane is ready to take off. These habits are codified in checklists. There are checklists for all kinds of standard operations, but they are most important during the most dangerous phases of flight: landing and take-off. These checklists are often long, and a short-haul pilot may have to go through them up to ten times a day.

In the Flight 1141 investigation, the focus was on the critical section where the Second Officer called out 'Flaps' and the First Officer replied, 'Fifteen, fifteen, green light.' This was the correct checklist and the response was also correct. If the First Officer had really checked that the flaps and slats were set, the accident would not have happened. But the gap between the call-out and the reply was less than one second, which, the NTSB concluded, was hardly long enough to perform the relevant check. Could it be that flight-deck habits designed to increase safety had become so automatic that it was ending up endangering safety?

Subsequently, human-interaction experts carried out interviews with pilots working at seven major US carriers.[2] They found that, because of the endless repetition, pilots quickly got into the habit of running through the checklists, and, once practised, did so automatically, often without really checking. Pilots reported that sometimes they only saw what they expected to see, rather than what was really there. They got into the habit of seeing the control in the right setting so that even when it was wrong their habitual perception of the situation overrode reality. When pilots went through the checklists, there was sometimes a sing-song quality to it. Instead of looking at the instruments and switches they were supposed to be checking, all of the replies were done from memory. This seemed to be what had happened on Delta Air Lines Flight 1141, which led to the mistaken conclusion that the plane was correctly configured for take-off when, sadly, it wasn't.

Accidents like this are by far and away the exception and are getting rarer all the time. By and large, routine checklist procedures are what make air travel so safe (that, along with computer fail-safes to pick up pilot errors). Indeed, much of the research on checklists has been carried out in the airline industry because small, routine mistakes can have such devastating consequences. In modern aircraft, checklists have now moved from paper to computer, with a corresponding reduction in errors made by pilots of almost 50%.[3] Pilots don't just complete checklists for the aircraft, they also run checklists on themselves. The so-called IM SAFE checklist goes like this: Illness, Medication, Stress, Alcohol, Fatigue/Food, Emotion. While it doesn't matter to other people that much if we're in a really bad mood at work, for a pilot it could be deadly for everyone on board.

The challenge for the airline industry is in standardizing habits and of working out why they fail when they do, because everyone's safety relies on it. The smallest slip has the potential to turn into a major disaster. For the majority of us, though, everyday slips in our habits seem less crucial. We continue to put on the same shoe first, visit the same coffee shop, and (at least some of the time) think the same thoughts. The reason that slips and mistakes in habits are so important is the same reason doctors have learnt so much about the human body from the study of disease: these are opportunities to learn how the system works, should we so choose. Why do our habits operate some of the time, but fail us at other times? What is it that ensures the routine performance of a critical habit? The anatomy of a habit is revealed in the way it fails, perhaps more so than when it succeeds.

~~~

On 6 March 1989, train driver Joseph McCafferty pulled out of Bellgrove Station in Glasgow, Scotland at 12:47pm. The train was a regular commuter service headed for Airdrie, a town 12 miles from Glasgow. He never made it. The train only travelled about half a mile before it ploughed head-on into a train travelling in the opposite direction. The driver of the other train and a passenger were killed. Driver McCafferty had to be cut free from the wreckage and lost a leg in the accident.

When the UK Railway Inspectorate investigated, it emerged that Mr McCafferty had pulled the train out of the station while the signal was at red, indicating that it wasn't safe to proceed. How had this happened? Yes, he was a young man of 22 when the accident happened; yes, he had only been a fully-fledged driver for about 5 months. But still, he was fresh out of a good training programme which was deemed satisfactory by investigators, so what had gone wrong?
~~~

The guard on Driver McCafferty's train that day was Robert Bain, who had nine years' experience in the job. It is the guard's responsibility to check that all the passengers are either on or off the train, and that the signal on the station indicates that it is safe for the train to proceed. He later admitted that he had not checked the signal, partly because it wasn't easy from his position at the back of the train and he knew the driver would be able to see it clearly from the front. So, quite routinely, he gave the ready-to-start signal to the driver, which consists of two rings on a bell. The driver gave two bells in acknowledgement and started to ease the train out of the station. He later told the inquiry that he thought the signal had been at green when he entered the station, and didn't see any reason to check whether it had changed. The signalman, however, stated that the signal had never been at green, and would have shown red during the whole time the driver would have been able to see it. Worse for the driver, even after pulling away the red signal would then have been visible to him for another 13 or 14 seconds, but he still didn't notice it, being more concerned with the speed of the train.

In the final report, McCafferty, the driver, got the majority of the blame for the accident, with the guard cited as a contributory factor. But ultimately, it is the driver's responsibility to check that it is safe to proceed, not the guard's. Psychologically, what had happened was that the driver had built up a simple habit. When he heard the 'ding-ding', he acknowledged it and set off without checking the signal himself.

This type of accident was eventually tackled with a simple environmental change. A switch was put in the driver's cab called the 'Driver's Reminder Appliance'. All it does is cut power to the train when it is activated. Drivers turn it on when they stop at a station as an extra safety check. If they get the ding-ding and try to apply power immediately, the train won't budge. They have to turn off the reminder switch, and this prompts them to check the signal first. Even better are systems which automatically stop the driver passing through a red signal by activating the brakes.

~~~~~~~

What happened to Driver McCafferty and the Delta pilots was tragic, but also incredibly routine: they didn't involve very unusual circumstances, or much that was out of the ordinary. Both were very basic mistakes with tragic consequences. In both cases, habits let people down, although in different ways. The pilots thought they had carried out a routine operation – checking the flaps and slats – which, actually, they hadn't, while the train driver had carried out a routine operation – pulling away from the station – but he'd done it at the wrong time, when the signal was at red.

Psychologists call what happened to Driver McCafferty a 'slip': when you carry out an action which you don't intend. As he found out, whether habits are good or bad, under certain circumstances they are very difficult to avoid. Stress, in particular, drives us towards our usual, habitual response to a situation, even when we plan to do otherwise. The extent of this can be seen in a study in which participants were
~~~~~~~

asked to learn (on paper) routes through a fictional underground rail system.[4] First, they practised the routes for a time, building up the habit of how to travel from one place to the other, just as we do for any regular journey. Then, half were told the underground map would change and that they would have to work out a new route. When participants were put under only mild time pressure, they only reverted to routine 30% of the time. But when the pressure was really put on, participants reverted to routine 70% of the time.

Habits cause us all to slip up some of the time, although usually without serious consequences. The psychologist and usability engineer Donald Norman has examined thousands of human slips made in all sorts of circumstances.[5] These include analysis of pilot errors and official government accident reports, as well as everyday slips in language. In one large class of slips, a habit is activated at the wrong time because it is brought to mind by something in the environment. William James, the nineteenth-century American psychologist who wrote brilliantly about habits, described many of these sorts of everyday slips over a century ago.[6] He wrote about a man going into his bedroom to dress for dinner (back in the days when people dressed for dinner) who got distracted and ended up getting into bed. Normal calls this a capture error: when the habit associated with a particular situation is so powerful that it overcomes all else. I'm in the bedroom, I've forgotten exactly why I came in here, so I might as well get ready for bed.

These sorts of errors are so ubiquitous, I've just made one

myself. Perhaps you spotted it? At the end of the last paragraph, instead of writing 'Norman', I wrote 'Normal'. 'Normal' is such a common word that it overrode 'Norman' automatically. I only spotted it after rereading the paragraph. If you didn't spot it, well, that's also perfectly norman (I'll stop now). That's because we also get into the habit of seeing what we expect to see, rather than what's really there, just like the First Officer on the tragic Delta Airlines flight. This has been tested in less dangerous circumstances by John Sloboda, an expert on the psychology of music. He gave musicians musical scores containing intentional mistakes and asked them to play from them.[7] What he found was that the musicians corrected many of the mistakes automatically by inferring the correct notes from experience. In the same study, he also found that people don't notice spelling mistakes, especially if they are in the middle of words. When it's on autopilot, the mind sees what it wants to see, or at least what it expects to see.

Many routine slips have a comedic quality, and Donald Norman lists many everyday examples of the kind we've all made from time to time. One involves a person thinking about asking another to make more coffee, but forgetting to actually say anything. After noticing it hasn't arrived, they complain. It only then becomes obvious that the thought has remained unspoken. This class of slips – when an action is substituted with only a thought – is clearly one of the most difficult to detect. We probably make these sorts of mistakes much more often than we know. A lot of the time, the consequences are so minor it makes no difference: we forget to pick up milk on the way home, or to charge up the laptop. In Norman's research, many slips were only detected with the help of an observer and sometimes, even then, neither actor nor observer noticed that mistakes were being made. Once again, in the psychology of mistakes, we see the unconscious nature of habits making them difficult to control.

Different types of amusing slip-ups also happen when habits get activated at the wrong time or are interrupted halfway through. James Reason, a psychologist and expert on human error, calls this 'the price of automization'. He reports people trying to pour a second batch of boiled water into a teapot with no recollection of already having filled the pot, and people putting second cigarettes into their mouths without realizing they haven't lit the first one.[8] Similarly, Donald Norman reports a story about a student who was queuing at a salad bar with a five-dollar bill in hand. A few croutons fell off the plate onto the tray and, in the attempt to eat them, both hands came up at once, but the croutons ended up on the tray and the five-dollar bill in his mouth.

These types of mistakes are most common when we're in familiar circumstances.[9] We make them in our bathrooms, kitchens, bedrooms, offices, and so on. It's why we go to the bathroom intending to brush our teeth and end up brushing our hair. Or why we go to the fridge to get milk and end up with a glass of orange juice. It's not just that these are the places we are most likely spend the most time; it's that these environments are chock-full of cues for our habitual behaviours, just waiting to be activated as soon as we get distracted.

~~~~~~~~~

The phrase 'health and safety' now strikes fear into our hearts. It has picked up all sorts of negative connotations: we think of needless caution, killjoys, fluorescent jackets, and packets of peanuts which have written on the side: 'may contain nuts'. But behind this smokescreen of journalistic clichés are people trying to work out how to prevent accidental injuries and deaths. Until the 1970s, these kinds of people relied on what seems like a very sensible theory: they assumed that people would respond to education. For example, if you want people to wear helmets when riding motorbikes, you educate them. You tell them about brain injuries, compound fractures, wet roads, and blind spots, and then let them choose to wear helmets. Or, if you want car drivers to wear seat belts, you show them films of people being launched head-first through their own windscreens. Who wouldn't respond to these sorts of shock tactics? Well, it turns out that quite a lot of people don't respond.

In the UK, for example, from 1972 the government spent millions on ads about seat belts on TV, radio, and billboards. It's estimated that 80 to 90% of British people saw these ads 8 or 9 times each.[10] Still, after all this effort to change their behaviour, only about 20 to 40% of people were wearing seat belts. The student of habits can tell you immediately that thc educational method isn't going to work and why. We've seen that psychologists can change people's intentions, but this frequently fails to translate into any change
~~~~~~~~~

in their behaviour when faced by strong habits. It wasn't until the law was finally changed, in 1983 – along with strict policing – that people's behaviour really shifted, with rates shooting up to around 90% or more.

Around the 1970s, there was a quiet revolution in how we thought about preventing injuries.[11] Some of the experts started to see that people didn't take much notice of attempts to educate them: in fact, they carried on doing exactly what they'd always been doing. They began developing models of injury prevention which stressed the interaction of people with their environments.[12] Instead of seeing us as completely autonomous, individual agents, separate from the environment, injury-prevention experts started to see the web of situations we are caught in. We aren't just people walking around, doing exactly what we want, driven only by our own motivations; we are embedded in our physical and social environment, responding partly automatically.

It has been argued that our individual psychology and biology is just the tip of a metaphorical 'injury iceberg'.[13] These are the parts of the iceberg that sit above the waterline, the bits we can see. Then, below the waterline, are all the causes of accidents which aren't so obvious: our homes, our workplaces or the organization we're working for, our social class or other aspects of our community. At the deepest level is our place in society and our national psyche. Habits are a result of all these levels interacting. We do what we habitually do not just for purely individual reasons, but because of this iceberg of interrelating and mostly hidden factors. What injury-prevention experts began to notice was that their efforts couldn't just target the individual; they needed to target the environment.

These shifts in thinking within injury-prevention circles were mirrored in psychology. Through the 1960s, 70s, and 80s, the way many psychologists thought about behaviour change was challenged. It emerged that getting people to stop smoking, do some exercise, or make any change at all was much harder than had been thought. The assumption that people could just will a change and it would happen was questioned. The models psychologists were using tended to be good at predicting how people behaved in one-shot situations, like when people in a lab are told about the health benefits of fruit and then given a choice between fruit or cake. But these models didn't work so well in the real world.

Back in injury prevention, the buzz was all about changing contexts. If you could change the contexts in which people made decisions, then you had a better chance of changing their behaviour, especially if that behaviour was highly habitual. Like the effort to encourage people to wear seat belts, only telling them that cigarettes were unhealthy did little to change their behaviour. But when laws started to restrict where people could smoke, this had an effect.[14] This was a change in the legal and, therefore, social environment, which was highly effective in getting people to change long-standing habits.

Another environmental solution for regulating human behaviour in dangerous situations is checklists. Although we've already seen how these can go wrong, their introduction did increase safety in aviation. Still, it has taken a long time because the change has had to reach right down to the bottom of the 'injury iceberg', trying to influence the way pilots behave through their peer group, their training and education, even their very psyche. Unfortunately, the lessons learnt from aviation have only spread so far. Product manufacturing is one area which knows all about the dangers of relying on human habits. Both food and car manufacturing have stringent monitoring procedures in place. Perhaps most enthusiastic in their use of checklists, though, are agencies which monitor the manufacture of medical devices and pharmaceuticals.

One area ripe for colonization by checklists is healthcare, a point put forcefully by surgeon Atul Gawande in his excellent book *The Checklist Manifesto*.[15] Just as bad or poorly implemented habits can kill, so good and well-implemented habits can save lives. One study of the US healthcare system has found that, each year, there are 1.5 million preventable injuries from medication errors alone.[16] How many of these are related to habits is unknown, but studies have shown that checklists can help because people's habits can break down in pressurized situations, or simply through boredom. An Australian physician, Dr Alan Wolff, has found that the introduction of checklists at one hospital increased staff compliance with key medical procedures by up to 50%.[17] Other studies have found that they can improve patient safety in many areas, including anaesthesiology and intensive care.[18] These are just a few examples; there are many more.[19]

In some ways, this focus on the environment in injury

prevention seems obvious, but actually we routinely make the same mistakes when thinking about ourselves. We think of ourselves as biology, psychology, and behaviour, and neglect the deeper levels of ourselves: how we are embedded in our environments, both physical and social. Our habits, just like the other parts of our thinking and behaviour, also grow out of these. The situation isn't just important for airline pilots, train drivers, or public health researchers; it's important for all of us battling against our everyday bad habits. The lesson that's been learnt in all these fields is the same lesson we have to learn ourselves: the situation has more power to control our habits than we like to think.

8

ONLINE ALL THE TIME

We don't know her name, but her problem illustrates a new fear. According to the short case report in an academic journal, a 24-year-old woman presented herself to a psychiatric clinic in Athens, Greece.[1] She had joined Facebook eight months previously, and since then, her life had taken a nosedive. She told doctors she had 400 online friends and spent five hours a day on her Facebook page. She recently lost her job as a waitress because she kept sneaking out to visit a nearby Internet café. She wasn't sleeping properly and was feeling anxious. As though to underline the problem, during the clinical interview she took out her mobile phone and tried to check her Facebook page.

Is this the fate that awaits us all once the machines take

over? Brains zapped into a robotic, unthinking, servile state by our online compulsions? Certainly, many of us have already caught the Internet bug: we know how useful it can be, and we wonder how anyone lived before they had instant access to such a huge and transformative resource. In ten years, we've moved from a situation where the Internet was mostly the preserve of geeks and nerds to the current situation, where it is mainstream. At the same time, we are nervous about what the Internet is doing to our minds. We worry about the increasing amount of time we spend using computing devices, whether laptops, phones, or tablets. We worry that the Internet is making us lazy, destroying our ability to concentrate, and sucking up time we would otherwise be spending in the real world. We love the Internet, but we also hate it. We want to use it, but we're sure we need time away from it. It's this ambivalence about our Internet habits that draws so much interest.

~~~

The Internet's original killer application is, of course, email, that spam-spewing slave-driver we love and hate in equal measure. Only a quick glance online reveals control over email-checking habits is a widespread concern. Our email habits certainly produce some frightening statistics. How often do you check your email? It's probably more often than you think. One study has found that people set their email program to check automatically every five minutes,[2] while an AOL survey reported that 59% of respondents check their email from the lavatory – that seems like taking
~~~

multitasking to extremes.[3] This habitual email-checking may eat up as much as a quarter of our working day, maybe more.[4] Perhaps part of the reason that people report email being so disruptive is that it can take over a minute to 're-cover' from the interruption and carry on with the task you were previously carrying out.

Email is incredibly habit-forming because the way it arrives is similar to what behavioural psychologists call a variable-interval reinforcement schedule. The 'variable interval' refers to the fact that if you check your email every five minutes, then most of the time you won't get anything and you don't know how long it will be until the next one arrives. The same is true of the emails themselves. Some are much more interesting and exciting than others. An email from your sister is likely to capture your attention more strongly than spam from an office supply company.

What really needs explaining is why we check our email so obsessively when so much of it is complete rubbish. Often, it's spam or a boring email from the bank to tell us about a tiny change to the terms and conditions. Let's imagine a ridiculous situation to show why we keep checking. Let's say one interesting email is guaranteed to arrive each hour, on the hour. In this case, you would only check then. If one time the gossip, or special-offer-too-good-to-refuse didn't arrive, then you'd be sorely tempted not to bother checking again. And, if there was still no juicy tittle-tattle or offer to become CEO of a major multinational a further hour later, you'd be even less likely to keep checking. When the expected schedule is off, we experience frustration, like when a train or bus fails to show at the appointed time.

That example sounds ridiculous because we know only too well how email really works. To see why email is so insidious, let's switch to a rat's-eye view. There's the lever in front of him, which he presses, and out comes a pellet. Success! So he hammers the lever a bit, but gets nothing for a while, then suddenly gets two, one straight after the other. Then, for a while, pellets are nowhere to be seen. Our experimental rat is getting a little frustrated but he learns that you never know, you might get a pellet or you might get nothing for ages. What happens is, he learns to put up with the frustration and settles down into a relatively slow habit of pressing the lever.

In the technical language of behavioural psychology, our rat is experiencing the 'partial reinforcement extinction effect'. In other words, he is not that bothered about failing to get a pellet when he presses the lever because he's used to it. The same is true of the human emailer. We get used to not being rewarded for checking our email most of the time; what keeps us going is that one time in fifty when something really interesting comes through. This combination of learning to accept non-reward along with occasional unexpected and unpredictable rewards is partly what makes checking your email so habit-forming.

There's a parallel here with looking for three cherries on a slot machine. Unlike email, though, which has a relatively random interval between rewards, slot machines pay out on a variable-ratio interval: there's some standard time between payouts, and this is varied a bit. The gambler then knows the machine will spew out coins at reasonably regular intervals, but isn't quite sure when – although, of course, it will always take more of your money, on average, than it gives back. Casinos know from experience that this type of variable-ratio interval is extremely powerful and will keep gamblers feeding in coins hour after hour. If you think you're addicted to email already, just be thankful the gambling industry wasn't involved, or you'd never look away from the screen.

Still, the email-checking habit is incredibly difficult to change, partly because it's so easy to perform, and, like many other habits, we do it almost without thinking. For some people, it's like a nervous tic or a time-filler in between other activities. These habits may irritate us, but they're not that damaging. The danger comes when checking our email starts to affect work or interrupt other, more enjoyable, activities. When you go away on holiday, the temptation to check work email can be huge. You resist a few times, but the idea keeps popping into your head that something might need attending to. And because you're on a regular schedule of email-checking, you are soon on your phone or laptop without thinking. In the back of your mind, you know it's a bad idea because you'll ruin the holiday thinking about work, but the habit is often so strong, it takes over.

For most of us, the biggest problem is the torrent of work email. Researchers have interviewed people in different professions about their strategies for dealing with email.[5] Three broad categories emerged. In the first were scheduled emailers; in this category, participants reported that they scheduled their email checking and kept it to once or twice a day, but no more. These participants were determined that email shouldn't rule their lives. In the second category were the obsessive emailers. One participant checked his email every 30 seconds and always responded immediately to every one that came in. Others freely admitted they checked their email far too often, constantly using it for both work and family matters, and many feared their mailbox would get too big. The last category were the deleters. These people simply deleted messages that were from strangers or people they didn't want to hear from – well, that's one way of dealing with it!

For those of us who have large quantities of email to get through at work, the aim is to become a scheduled checker, but where the schedule is a lot less than every 30 seconds. It will depend on how long you are able to concentrate on a particular task, but every 45 minutes is a reasonable compromise – the idea being that after 45 minutes spent concentrating on a task, it's about time for a break. Any attempt to change established habits has to take into account the situation. We're not rats running round cages, pressing levers; we're human beings with a broad range of choices. There are alternatives to checking your email, like resting your eyes or making a hot drink, but these are not so frequently and insidiously reinforced as exciting emails. If we want to change our email-checking habit, the research suggests that it needs to be replaced with another type of behaviour that is more attractive. In the language of behavioural psychology, we need to reinforce a different behaviour. Instead of checking email, we could give ourselves a 30-second 'screen break'.

For people who sit in front of monitors all day long, allowing the eyes to relax is a much better use of time than checking email. What's more, this space can be used to reflect on what you're working on.

If this doesn't work, then why not try a standard strategy from behavioural psychology: punishment. But since you probably don't want to rig your mouse to give you an electric shock when you try to check your email, we'll have to use the higher faculties of reason and understanding. First, consider the dangers of multitasking. Although it may feel as if we are able to do two tasks at once – say, talking on the phone and checking our email – what's actually happening is that our brain is quickly switching from one task to the other. Unfortunately, switching between tasks has costs: most notably, we are likely to forget what we were doing. Our intentions are surprisingly fragile, despite the feeling that we could never forget them. A delay as short as 10 or 15 seconds – the time it might take to read an email – is enough to put us off track.[6] When we come back from an email, it takes time to remember what we were doing, or we may not remember it at all. There are varying figures for how long it takes to recover from an email, but one study suggests 64 seconds.[7] Another study suggests people spend as little as 3 minutes on each task before they switch to another.[8] Not all of these switches were caused by email, but many were. And at each switch there is a similar cost to concentration and to the psychological state of flow: that feeling of being in the zone, in which we often do our best work.

Nowadays, email is ancient history, and the buzz is all about social media. Twitter in particular has captured people's attention. Twitter is a cross between a social network and a blog. The blog part is that users read and write 140-character 'tweets', which are largely public. The social network part is that people 'follow' each other, then become part of each other's Twitter conversations; they can also 'retweet', or retransmit, other people's messages to their own followers.

Although we are often encouraged to think of services like Twitter as brand new, 'bleeding edge' technologies, the fascinating thing about social media is that, conceptually, there's hardly anything new about it. When you send a tweet, it's like sending an email, except on Twitter it's limited to 140 characters and you cc all your followers and, effectively, the whole Internet. Behind the hype, we are still sitting in front of our computers or using our mobile phones to communicate information to other people.

Using Twitter, though, has the potential to be more immersive than email. With email, you generally have one conversation at a time; on Twitter, you can be carrying out many conversations all at the same time. Email is reminiscent of letter-writing, while Twitter is more like going to a virtual party. Messages from all the people that you follow scroll up the screen and you add your own messages to a huge public conversation.

The social aspects of Twitter, though, shouldn't be overstated. Everyone isn't talking to each other equally. Just like at a party, some people talk louder and longer, and a lot of people only talk about themselves. In one piece of research which analysed over 350 Twitter users, only 20% shared information with other users and replied to them: this group were dubbed 'informers'.[9] On the other hand, 80% only broadcast information about themselves to their followers: they were dubbed 'meformers'. The meformers tended to have smaller social networks, perhaps because they were less likely to pass on pieces of information that would be of general interest and give a hit of pleasure to their followers.

More evidence that Twitter usage is lopsided comes from a study of 300,542 users in 2009.[10] This found that just 10% of the users generated 90% of the tweets. Many of the most popular tweeters were, unsurprisingly, celebrities. The other 90% sent out little or nothing. The median number of tweets over the full lifetime of the account was exactly... 1. In other words, most people probably only sent out their first tweet (if that) when they joined and then never posted another message. Since that 2009 study, it is possible that people are already posting to Twitter more than they used to. Still, it is likely that while the average user's output will pick up, the overall pattern will remain the same, with the vast majority of users listening attentively to a minority.

Although people expect the social elements of Twitter to be most gratifying, in reality, when they start using it, the informational aspects are the ones that prove most satisfying.[11] Most Twitter users think of it more as a useful information source than as a social network. An analysis of the networks and paths of tweets that travel between people supports this, with less than a quarter of Twitter users having reciprocal links between them. This is where the analogy of a party starts to break down: even the quietest and most withdrawn partygoers say hello to the host when they arrive, and manage to chat with a few people. Twitter doesn't seem to be like that. The average Twitter user is mainly waiting and hoping for a new piece of juicy information to come through.

This is where Twitter and email are similar: for many people, it's more fun to receive an interesting email than it is to write one. Like email, on Twitter an interesting tweet could arrive at any moment, but you don't know when. You could get a batch of interesting tweets one after the other, or nothing for a few hours. Once again, it's a variable-interval reinforcer, in which users get used to the frustration of not getting anything interesting for a while but keep checking anyway.

People join Twitter for all sorts of reasons: their friends are on it; they've heard about it in the media, or want to follow a celebrity. In some ways, the reason doesn't matter. If people derive some satisfaction or gratification from it early on, then a habit soon starts to form.[12] With habit comes the automatic, unconscious compulsion to check Twitter. What we see in Twitter usage is what we see in many other types of habits. At first, we decide on our actions deliberately and for specific reasons. Once the habit has established itself, our conscious thought processes fade into the background and Twitter use becomes self-sustaining. We check email because we check email and we check Twitter because we check Twitter. It's then up to the conscious, rational parts of our minds to check up on these habits every now and then and ask ourselves whether we are still getting something out of the behaviour or whether it needs reining in.

We've looked at both email and Twitter as two examples of how our online habits operate – email because it is the Internet's original killer application, and Twitter because it is one of the new kids on the block, and very popular. But really, they're just different forms of medium: ways of storing and delivering information. It just so happens that many of the media which are currently so fashionable are accessed through the Internet. The Internet is special and new in some ways, but in others it's just performing the same function as printing presses and radio waves and TV signals: it propagates information. And wherever there is information being propagated, there will be people building up media habits.

Take television as an example of an old-school media. Here's a simple question: why do you watch TV? Many people's answer to this question has something to do with uses and gratifications. You get something from it whether it's entertainment, education, or information. But lurking in the back of your mind is another suspicion, a suspicion that this book is hopefully doing something to cultivate. Perhaps watching TV has more to do with habit and passing the time than anything you actually get from it. Admittedly, passing the time is a kind of use, but not a very good one; it's the kind of explanation we give when we don't really know why we're doing something.

For decades, researchers have supported the idea of people as mostly consciously choosing how much they watch television and what they watch, but this view has been attacked by some who find evidence that our habits win out over our intentions.[13] We have a tendency to resist this explanation because it threatens our sense of being in control of ourselves (how dare psychologists tell me I don't know why I'm watching TV). But since people have relatively little insight into their internal cognitive processes, they can easily make up explanations for their habitual behaviour. None of this is to say that some television doesn't entertain, inform, and educate; clearly, some of it does. The point is that, like email and Twitter, what once provided us with gratification can easily become a mindless, automatic process which we explain away to ourselves with reasons that are long out of date. The key is to notice these habits, bring the decision back into the conscious mind, and try to spot when it's time to switch off.

~~~

Finally, we reach a word that may have been in the back of your mind throughout: addiction. If you can't stop checking your email account or Twitter feed, or, indeed, can't stop watching reruns of *Scrubs*, then are you addicted? Can a bad habit like this spiral out of control into an addiction? 'Addiction' is an emotive word and there is a huge debate going on about whether or not we can say that some people are addicted to the Internet. Some say it is impossible to be addicted to the Internet because it is a medium of communication. For
~~~

example, someone may spend hours online every day trying to find the cheapest price for vodka, but you'd be hard pressed to argue that they were an Internet addict rather than an alcoholic. This argument is true but misses the main point. Whether or not people can really be said to be addicted to the Internet, there are certainly bad habits to be explained. It's easy to be unaware of just how much we check our email or log on to Twitter. Switching from writing an important report to checking email or Twitter doesn't feel as if it interferes with report-writing; instead, we rationalize it as a break. But it may actually be a robotic, habitual behaviour that serves no purpose, not even giving us any pleasure.

Rather than talking about addiction, though, some psychologists have suggested that we should think about 'deficient self-regulation': we are out of control. There are two aspects to problems with self-regulation.[14] The first is deficient self-observation: we don't notice how much time and energy we are putting into online activities. The second is deficient self-reaction: we don't seem to have control. Each feeds into the other, making it less likely that we'll be able to make a change.

Whatever it's called, in the most extreme cases people do seem to be using the Internet in a way that looks like an addiction. Addicts become heavily preoccupied with it, getting arousal while using it and experiencing cravings for it while away.[15] They find they need to use it more and more to get the same effect. They start to lose control of their habit and feel withdrawal symptoms when away from it. The habit starts to eat away at other activities, close relationships, work, and anything approaching a normal life. Performing the habit becomes a method of escape. At the same time, addicts try to conceal their addiction from others.

While these problems do exist, it is likely they've been overstated in the population as a whole. Much of what we know about how people use the Internet has come from studies in which the participants are already high users of the Internet, such as the kind of people who respond to online adverts for study participants. Some have found that 80% of participants are dependent on the Internet, spending an average of 38.5 hours on it a week.[16] But this is a bit like judging how much alcohol people drink on average by heading to a bar at happy hour. When you extrapolate from people who are deeply immersed in the Internet, it does look as if the machines are in control. When you recruit people across a range of social classes and age groups, you get a much less frightening answer. One Swedish study from 2011 surveyed over 1,000 people and found that the weekly average amount of time spent on the Internet at home was 10 hours.[17] The vast majority of people in that study didn't use the Internet excessively, with only 5% using it more than 30 hours a week.

Regular media habits don't become really bad media habits or addictions without some help – often, depression or anxiety is involved. For those experiencing these states, being online starts as a pleasant distraction from everyday worries but can soon become a problem of its own. What can happen is that going online becomes a habitual reaction to feeling depressed, anxious, or experiencing some other aversive state. Once the linkage between emotional state and habitual response is established, it can be remarkably difficult to break down. It's a vicious circle because depression leads to bad media habits, which leads to negative life events, which then leads back to more bad media habits. It's like the Greek woman in the case report at the start of the chapter: her Facebook use meant she lost her job, which led to more time on Facebook. The stronger the habit gets, the worse life gets, and so the more attractive it becomes to escape into the screen.

This talk of addiction might seem a bit much if all you're worried about is checking your email or Twitter feed too often. And, of course, for the vast majority of people these symptoms will never be experienced to these extreme levels. But for anyone who's ever felt some aspects of online activity start to take over their lives, a few of these symptoms will be familiar. People talk about withdrawal when away from their email, checking it secretly and feeling ashamed. At the extreme, it can push out other activities and feel like it's getting out of control.

These behaviours can easily continue at work where the fairly normal activity of cyber-slacking – shopping, gaming, or blogging on company time – can quickly turn into cyber-absenteeism. A survey of office workers has found that they spend an hour each day on non-work-related activities, of which almost half were online.[18] This is probably a conservative estimate which would increase dramatically if the Internet logs were examined. The fact that so many of us are sitting in front of computers all day at work and then again in the evening – and on Internet-enabled mobile phones in between – means that the opportunity is always there for our online habits to get out of control.

~~~

Email and Twitter are great examples of how popular online activities can be habit-forming, but these same mechanisms and processes operate in other online and offline media to greater or lesser degrees. For example, online gaming may be more addictive than Twitter, Facebook, or any other social network, but the experience can hook users for just the same reasons. Games reward us with the pleasure of finishing off the bad guy, solving a puzzle, or getting to the end of the level, while Twitter, Facebook, TV, and the rest reward us with little bits of information, social connections, and entertainment.

The reason we're online all the time is that the rewards for certain online behaviours are so easy to learn. You don't have to move from your chair; everything comes to you. Because the association is so clear and obvious and happens so quickly, like the rat getting its pellet of food, the habit is easy to learn. As a result, it's no surprise that computers, phones, and other electronic devices are super-habit-forming in the way that previous non-interactive media weren't. They react instantly and often attractively to what we do. Watch other people tapping away at their mobiles or clicking on their mice, and what do they look like? Rats hitting a lever for a pellet? In many ways, that's what we are. Finally, it's the sheer ubiquity of the Internet that makes the services we can access through it so habit-forming. Our on-
~~~

line activities easily permeate almost all aspects of our lives. Potentially we can be online, without any effort, from the moment we wake up in the morning to the second we go to sleep. A tablet computer is by the bed, along with Internet-enabled mobile phone; at work, many of us are permanently online; in the evening, while watching TV, Google will tell us the actors' heights and ages; while cooking, we can check Twitter; while falling asleep, we can listen to Internet radio.

Technology is creating a world where people can and do fill up every moment of time with online activities of one kind or another. Our online habits can be layered one on top of the other so that there is no time for anything else. Studies of the ways in which young people interact with computers show that they very rarely do only one thing at a time. More normally, teenagers are to be found watching TV, instant-messaging, surfing the Internet, and downloading music, all at the same time.[19]

The debate continues about what harm our online habits might be doing to us. Some studies suggest that heavy multitasking habits may make us less able to concentrate on one task and perform it properly.[20] But the research is at such an early stage that it's difficult to draw any solid conclusions. What we can say is that, like any technological advance, the Internet is not inherently good or bad in itself; it's how we learn to use it that matters. With all the advantages being online gives us, we are also offered a set of potential dangers we have to understand. What we know about how humans react to virtual environments is still in its infancy, but we can be sure we will be offered up new online services tailor-made to engage our habits. In the battle between intention and habit, we need to be able to work out who is winning: who is master and who is slave.

PART 3

HABIT CHANGE

9

MAKING HABITS

Years ago, when Jerry Seinfeld was still on the comedy circuit, a young comedian asked him for advice on how to improve.[1] Seinfeld replied that the key to being a better comedian was writing better jokes, and the way to write better jokes was to practise. But it wasn't just about practising in general, Seinfeld explained; it was about building up a habit – the writing habit. Seinfeld advised using a simple trick to get the habit going. You buy a big wall calendar with a box for every day of the year. Then, each day that you complete your writing task, you put a big cross on the calendar. As the weeks pass, the chain of crosses on the calendar grows longer and longer. Your only job, urged Seinfeld, was not to break the chain.

These kinds of stories contain a seed of truth about making a new habit, as do many snippets from the lives of successful people. It's why they're so fascinating: we sense that these daily routines hide some profound secret about how to achieve greatness. On closer inspection, we discover that their daily routines are often profoundly simple; with some honourable exceptions, people achieve great things by working steadily and regularly towards their goals (along with more than a dash of help from genes and circumstances).

Take one of the most influential scientific minds in history, Charles Darwin. According to his son, Darwin's routine was metronomic in his middle and later years.[2] He rose at 7am, breakfasted alone, then worked in his study from 8 until 9:30am, during which time he got his best work done. Then, he broke for an hour's letter-reading before returning to his study for a further hour-and-a-half's work. The rest of his day was taken up with walking, lunching, reading, letter-writing, and family matters.

Or take the great comic writer P. G. Wodehouse, creator of Jeeves and Wooster and the Blandings Castle novels. Wodehouse rose at 7:30am, carried out his 'daily dozen' calisthenic exercises, had breakfast (tea, toast, and coffee cake), went for a short walk which he had done for years, before settling down to write at about 9am until he broke for lunch at about 1pm.[3] The rest of the day was taken up with a long walk, tea and cucumber sandwiches at four with his wife, possibly followed by some more work, then a lethal martini at six, and the rest of the evening was spent reading or playing cards with his wife.

While the habits of successful individuals are interesting, they often aren't that practical or easily applicable to the habit you want to create. Neither of these spare timetables tells us how Darwin came up with the theory of natural selection, or how Wodehouse managed to get Bertie Wooster into and then out of so many entertaining scrapes (although both would have been impossible without the work habits they'd established). Neither does Seinfeld's productivity tip explain his brilliance – nor does he mean for it to. Successful people can provide inspiration and motivation, but not necessarily a blueprint for a new habit. As scientists are sometimes heard to mutter, the plural of anecdote isn't data. Their success stories provide fragments and seeds, but they don't provide the specifics. They can tell us about the physical processes, their timetables and foibles, but they aren't so good on their mental processes. We are left with all kinds of questions: What habit should we try to create? Where does the motivation come from? When and how should you perform the behaviour you wish to become a habit? How should you address failure, dissatisfaction, and everyday inconveniences?

The following stories of success in making habits are from scientific studies. Hundreds, sometimes thousands, of people in these studies have been trying to make changes in their lives, and psychologists have measured how successful they've been and drawn conclusions about what methods work best. From these studies emerges a series of techniques that should be applicable to almost any type of habit.

Before we look at how to build a new habit, we need to take a step back to think about motivation. Why, exactly, do you want to make a new habit? Sometimes, the reasons are obvious and don't need any further soul-searching, but this isn't always the case. People often charge into trying to make new habits without asking themselves what the new habit will do for them. There has to be an ultimate goal that is really worth achieving, or the habit will be almost impossible to ingrain. What we find in the research is that when people's goals start to weaken, or are weak in the first place, it's very difficult to start forming a new habit. A few minutes spent thinking about this before you dive in will pay dividends in the long run.

Let's start by giving one piece of bad advice a knock on the head. Many popular self-help books tell you that imagining successfully completing your goal is beneficial. The theory goes that if we can picture our future success then this will help motivate us. There is some truth to the idea that being positive about the future can be beneficial, but there are also pitfalls. One of the main ones is fantasizing about reaching our goals, which can be dangerous. The bad news about fantasizing was underlined by a study that pitted it against positive expectations.[4] This research found that participants who expected success, rather than fantasizing about it, were more likely to take action. The problem with positive fantasies is that they allow us to anticipate success in the here and now. However, they don't alert us to the problems we are likely to face along the way and can leave us with less motivation – after all, it feels as if we've already reached our goal.

Expecting success is about being practical. We have to think carefully about what is really possible. One way of making it more obvious what kinds of habit change are possible is by using visualization. As opposed to fantasizing, a more effective way of visualizing the future is to think about the processes that are involved in reaching a goal, rather than just the end state of achieving it. One study that demonstrates this had students either visualize their ultimate goal of doing well in an exam or the steps they would take to reach that goal, in this case, studying.[5] The results were clear-cut. Participants who visualized themselves reading and gaining the required skills and knowledge spent longer studying and got better grades in the exam than those who only visualized their goal. One of the reasons just visualizing an outcome doesn't work is the planning fallacy. This is our completely normal assumption that reaching our goal will be much easier than it really will be. It still strikes people, even after years and years of experience. It can be difficult to anticipate just how much of any plan can (and will) go wrong.

A word of caution about choosing a new habit to establish. With habit change, people often try to bite off more than they can chew. One response to being unhappy with ourselves is to go for a complete reinvention – try to avoid this. Almost all the research on making new habits only addresses quite simple behaviours, and still some people regularly fail. It's much better to start small, and if the procedure works, then run it again for the other habits you want to establish. Alternatively, you can break down a larger habit into its component parts and work on each part separately.

The famous behaviourist B. F. Skinner used this method, known as 'shaping', of building up habits one by one to get pigeons playing ping-pong. If that doesn't impress you, then he also had a rat reacting to 'The Star-Spangled Banner' by hoisting a small American flag and saluting with its front leg. It's all about layering one simple habit on top of another.

Making habits, though, is about more than just process. People tend to have quite free-form ideas, bordering on fantasies, about how to change themselves; whereas what we need for our new habits to stick is a concrete goal to which we are committed. All those repetitions we need to carry out aren't going to happen without commitment. Psychological research has looked at the techniques which will help weed out the fantasies and consequently boost our chances of creating long-lasting changes in ourselves. Two techniques we might naturally use to make habit-change plans are to imagine the problems solved and to think about why we are unhappy with the current situation. In different ways, each might provide us with the motivation we need. But do they work?

In a series of experiments led by Gabriele Oettingen of New York University, these methods were pitted against each other, and a new technique was added. As a test, participants in the study were given a problem to solve and divided into three groups.[6] In each group, they were told to use one of three strategies while solving the problem:

1. *Indulge*: imagine a positive vision of the problem solved.

2. *Dwell*: think about the negative aspects of the current situation.
3. *Contrast*: this was the new technique. First, participants imagined a positive vision of the problem solved, then thought about the negative aspects of reality. With both in mind, participants were asked to carry out a 'reality check', comparing their fantasy with reality.

The results showed that the contrast technique was the most effective in encouraging people to make plans of action and in taking responsibility, but only when expectations of success were high. When expectations of solving the problem were low, those in the mental contrast condition made fewer plans and took less responsibility. The contrast condition appeared to be forcing people to decide whether their goal was really achievable. Then, if they expected to succeed, they committed to the goal; if not, they let it go. This is exactly what we are looking for when we want to make a new habit. We need to know as early as possible what we can commit to.

This all sounds fine in theory, but the main problem with mental contrasting is that, in practice, it's hard. Thinking about the negative aspects of our goals is unpleasant; similarly, bringing fantasy and reality together is uncomfortable because suddenly it becomes obvious what needs to be done, and these realizations can be depressing. Another reason the technique is difficult is that people dislike moving from happy to depressing thoughts. If we feel happy, we want to keep thinking about happy things, and if we're thinking negative thoughts, it's difficult to change to positive.

The clash between the fantasy of, say, developing a habit of practising the piano and the obstacles you'll face can be depressing. When you project yourself forwards, you know it will feel wonderful to play those Bach keyboard concertos perfectly, but how will you put in the hundreds or thousands of hours of practice required? How will you create the time and space in your life away from family and work? More importantly, have you bought a piano yet? Once you have faced the reality of how you'll find the time and determination to practise, though, the research suggests you'll take action sooner, feel more energized and make a greater emotional commitment to building the habit.

If you're finding the mental contrasting difficult, then try the 'WOOP' exercise.[7] WOOP stands for 'Wish', 'Outcome', 'Obstacle', and 'Plan'. First, you write down your Wish, the habit you want to achieve; then, the best Outcome of your habit; then, the Obstacle(s) you are likely to face. Finally, you make a specific type of Plan called an implementation intention, which is described in detail below.

Now that we've got a clear goal in mind, we are ready to start work on the new habit.

~~~

With your motivation identified, stage one of building up a new habit is to have a plan, but not the ordinary kind of plan we might automatically try to use. The regular types of plans that people make tend to be vague. We say things to
~~~

ourselves like: this year, I really want to get fitter or I want to be kinder to my partner. What these types of plans lack is the exact behaviour and the exact situation in which we will perform it. So instead of saying to ourselves 'I intend to get kinder or fitter', we should say, '*If* I see someone struggling with a buggy, *then* I will offer them help'; or, '*If* I'm about to get in the car for a short trip, *then* I should walk'. This links a particular situation with a response, an action. Recall that what we want is a strong linkage between a specific situation and an action – once this connection is automatic, we'll have a new habit. This implementation intention, this if-then link, is like an embryonic habit; it's the blueprint for the habit to come.

We know that implementation intentions are better than the plans people make spontaneously because this has been thoroughly investigated in many experiments. In one study, for example, participants were trying to increase their fruit and vegetable intake.[8] People in the control group made whatever plans they saw fit, whereas an experimental group made specific implementation intentions. After one week, the control group had managed absolutely no increase in fruit and vegetable intake at all. On the other hand, those who had used implementation intentions ate an average of an extra half-portion each day. This difference couldn't be explained by greater motivation in the experimental group, so it supports the idea that implementation intentions are effective. This is just one example; implementation intentions have been tested in study after study. People have been observed while trying to initiate all kinds of new habits, from exercise to doing puzzles, eating healthily, collecting coupons, recycling, and many more. Ninety-four of these studies, with a combined total of over 8,000 participants, when collected together found that the effect size for implementation intentions was medium to large.[9] This is psychologists' shoptalk for *it works*.

To work properly, though, implementation intentions have to be made correctly. We need to break down if-then implementation plans into their two component parts.[10] First, the 'if', which is the situation or trigger for your action. The question here is: how specific should you make the 'if'? When the 'if' is too specific, it could limit the possibilities for actions. For example, imagine you say, '*If* I reach the lift at work, *then* I will take the stairs'; this then limits you to the stairs at work. Very specific plans can also be too rigid: they can't take into account the vagaries of everyday life. Situations aren't always exactly the same and the plan needs to account for that as well. On the other hand, when the 'if' is too vague, it's easy to miss opportunities to practise your habit. The ideal 'if' component is balanced between the abstract and the specific: wide enough to include ample opportunities for practice, specific enough that it will trigger the vital second component, the action. In the lift example, a little tweak may fix it: '*If* I reach **any** lift, *then* I will take the stairs'. I say 'may' because implementation intentions, like many of the techniques I describe, require some trial and error. We know that they work in general, but they haven't been tested in every conceivable situation for every type of person. For example, let's say you're in the habit of taking the lift with other people. Are you really going to break off the conversation and take the stairs on your own? If not, then this may not be a behaviour that you can get to stick and turn into a habit. The implementation intention has to work for you and for the situations you find yourself in. They will almost certainly require some fine-tuning.

A final point about the 'if' component: timing. You might be tempted to choose a time of day; say, I'll go for a run at 8pm. Don't do this. The problem with using a time to cue up a new habit is that you have to be clock-watching.[11] And we can't be clock-watching all the time. It's far better to use an event. Events are much more likely to work because they don't rely on our memory, which, as you may have noticed, is notoriously unreliable. What you need is an event that can't be missed. Researchers have found that the best cue for a new habit is something that happens every day at a regular time. Participants in one study were trying to eat more healthily found that the cues of arriving at work and lunchtime worked well.[12] Effectively, they were adding on a new habit after one that was already operating. This kind of linkage is likely to work much better than using timing cues.

While setting specific times to perform habits is not recommended, it's very important to think about how a new habit will slot into your daily routine. New habits which aren't already linked to situations (like the lift/stairs example earlier) need to find a situation which will cue them. Think about how large portions of your day are habits linked together in chains.[13] What you want to do is add a new link in the chain where there is an open slot. You are looking for a time when you've just finished one regular habit and you're casting around for the next activity. Like putting out your rubbish after clearing the kitchen, or flossing your teeth after you brush. Look through your daily habits for an activity that forms the last link in a chain; then consider adding your new habit on here.

That's enough 'if'; let's have some 'then'. The main rule for the 'then' component of implementation intentions is that it should be specific. If you decide to be kinder to your partner, after selecting the trigger for your action, which might be when you are in the kitchen wondering what to cook for supper, you then choose a specific action: I will prepare their favourite meal. Generally, the simpler the 'then' component of the implementation intention, the easier it will be to carry out. But you can also specify more complicated actions, as long as those tasks themselves are automated. For example, driving a car to work is a complicated task, but for experienced drivers it's so automated that it counts as a simple task. The 'then' component doesn't have to be just one action. If you are planning to exercise more, but want to give yourself the option of different forms of exercise, then this can be beneficial. So you could specify: '*If* it's after breakfast and there is time, *then* I will go for a run or ride my bicycle'. A study in which people used this sort of multiple-option implementation intention showed that their task performance was improved.[14]

While implementations do work when given a chance, that doesn't mean they're foolproof – far from it. People who are naturally meticulous planners tend to benefit less from implementation intentions.[15] For the rest of us, both our own minds and the situations we naturally find ourselves in can easily tempt us off the straight and narrow. One way in which your mind will try to trip you up is by suggesting that all this effort you're going to isn't worth it. When performing your habit is hard and you're not strongly committed, you're likely to give up.

A second way in which our mind can sabotage us is through the vagaries of everyday moods and fears. You may have decided to start practising the piano in the evening, but after a hard day at work, you don't feel like it. You may have committed yourself to attending a dance class, but fear of looking like a fool puts you off actually going. How is it possible to overcome these aversive states? Again, we can use 'if-then' plans to shield ourselves from these attacks from the inside. We say to ourselves: '*If* I feel scared of the dance class, *then* I will remember that everyone is a beginner and scared of looking stupid'. Or: '*If* I feel too tired to practise the piano after work, *then* I will first listen to some inspirational music to help motivate me'.

These types of if-then plans for shielding your fragile new habit from adverse thoughts will require some work, but the effort is worth it. Their effectiveness has been tested in a situation in which many people suffer overwhelming, distracting, and self-limiting thoughts: while playing competitive sports. Over one hundred table-tennis players were split into three groups and given different instructions about how to play a match.[16] The first group were told nothing, the second group simply to concentrate as hard as they could, and a third were given a set of implementation intentions. This third group identified specific inner states that were problematic, such as '*If* I'm playing too defensively...' and then they chose appropriate responses, such as '...*then* I will risk something and play courageously'. Each player chose implementation intentions that were specific to themselves. Players who got angry tried to calm themselves, players who lost motivation tried to give themselves a boost, and so on. The results showed that those who used the implementation intentions improved their performance during match play compared with the other two groups.

It's not just our own mind that tries to trip up our new habits; it's also the situations themselves. These provide a whole new set of temptations to lure us away from our new habit. We go on a diet, and cakes, crisps, and beer are suddenly everywhere. Or we decide to walk to work, and then it starts raining every day. This is where 'coping planning' comes in. We know there are going to be temptations along the way and, with a little thought, we can anticipate what these are going to be. Much like we make implementation plans for our new habit, we also need to make implementation plans for situations in which we'll be tempted. This is exactly the same principle as the table-tennis players were using, but instead of targeting thoughts, coping planning targets situations. In one study of those trying to give up smoking, participants were asked when they were most likely to relapse.[17] The situations included after a meal, while having a drink with friends, and when under stress. They then formulated simple plans to enable them to cope with each of these situations. This proved beneficial to people trying to give up.

~~~

What seems like a fairly simple task of making a plan actually requires some creativity. Although people do spontaneously make plans, they aren't usually precise enough. We need these ifs and thens because of all the things that can go wrong. But with the right kind of planning you should be able to get your new habit off the ground. The next step is to keep it going.

Building a new habit means repeating the thought or behaviour in a stable context. Each time it is repeated, we go a little way towards increasing the habit's automaticity. Exactly how many repetitions are required will depend on your lifestyle and the exact habit you're trying to develop. Even the simplest variations can affect how many will be required. Recall the study from the first chapter which found a curved relationship between habit repetition and automaticity. Here's the graph again:

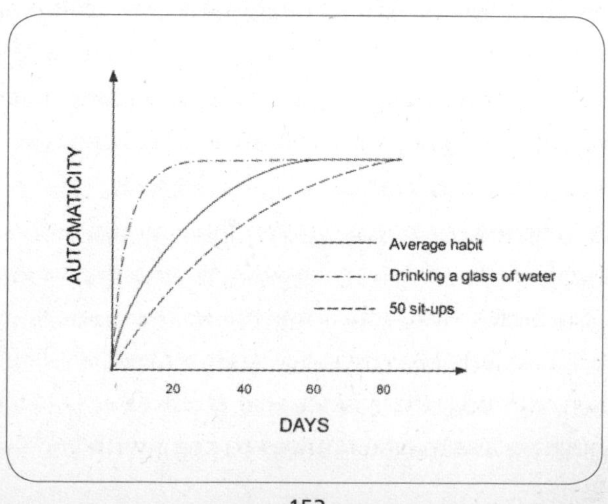
~~~

In that study, one person who decided to eat a piece of fruit with lunch took over 60 days to form the habit, while another person who ate a piece of fruit in the evening took only about 30 days. This difference may be down to personality and/or when it was most convenient for that individual to eat the piece of fruit. On average, though, the habits took 66 days to form. What the graph suggests is that each time you repeat your new behaviour, you take one small step up the slope towards a new habit. In fact, because the line is steepest at the start, the greatest gains are made with the early repetitions. Hopefully, the implementation planning will have started you on this upward slope.

Psychologists have looked at the factors that affect whether or not we keep going. As you'd imagine, satisfaction with the new habit is right up there. Consciously or otherwise, we ask ourselves whether our new habit is really how we imagined it. A feeling of satisfaction is like a message from our unconscious that we're going in the right direction, while dissatisfaction signals that something is wrong.

Dissatisfaction is a real killer for a new habit. People are often put off if they don't feel they're getting anywhere. There are all sorts of ways of coping with this: the best one for you depends on what you're most comfortable with. Some people prefer to get support from other people. To give an obvious example, joining a gym and attending structured, timetabled classes can be useful. On the other hand, it can be a colossal waste of money if the gym isn't for you. A cheaper way of getting support is to involve a friend or your partner in habit change. This will also give you a chance to talk about your dissatisfactions and how to address them.

For a more internally directed technique, implementation planning can be used again. When you start developing a new habit, ask yourself how you will cope with dissatisfaction. If you become dissatisfied, will you decide to give up, or will you decide to redouble your efforts? Studies have found that implementation intentions are vital to the continued performance of new habits, just as they are to getting them started.[18] Ideally, they should address the reason for your dissatisfaction. Although this is not always easy to determine, the usual suspects are a good bet: lack of progress, motivation, tiredness, and so on. Once again, the implementation intention you use should be directed at what you're feeling and should give you a way to get back on track. For example, a perceived lack of progress can be addressed by trying to think optimistically about how far you've come, rather than focusing on how far there is to go. Similarly, both mental tiredness and weakening motivation might be addressed by using music – try accompanying your workouts with a thumping soundtrack, or use a chilled-out selection to help you organize your latest batch of photographs. These are just a couple of suggestions; you will need to find what works for the particular habits you're trying to build.

An approach many people spontaneously use when starting up a new habit is monitoring progress. For psychologists, this doesn't just mean recording progress on a chart (although many people find this useful), but also monitoring the self during the day. It's about being aware of how your new habit is developing. Would it be better to perform it at a different time of day, or in a different way? What types of temptations do you feel to skip your new habit completely? Noticing problems or ways of improving can all lead to a habit that's easier to practise and so will become automatic more quickly.

It's more than just noticing, though; you have to act on the information. In one study of people trying to lose weight, some of the participants carefully monitored their eating, noting down both the temptations and distractions they experienced.[19] Despite this, they didn't notice how it could help them and so they didn't lose much weight. Once again, implementation intentions can be employed to help deal with problematic situations. The process of building up a habit should be iterative: you are learning what is easy for you and what is hard, what is possible and what isn't. If you are too harsh with yourself about the performance of a habit, your willpower is likely to give out. In particular, there is no need to be too harsh with yourself if you miss one or two opportunities to practise your new habit. Skipping the odd repetition won't hurt.

A final technique for starting up a new habit is both well known and laced with danger. It's rewards. The problem with rewards is that, ideally, we should be motivated internally rather than externally. Psychologists call this 'intrinsic motivation', and it's generally more powerful than external or 'extrinsic' motivation.[20] We want to avoid having our habit become conditional on some kind of reward. While rewards may work at first, we can easily become used to them over time and so they can lose their potency. For example, say you're working on a new habit of cleaning up the kitchen soon after eating, rather than leaving it until later. You decide to reward yourself with an hour reading a new novel that you're currently enjoying. The problems are: what happens when you get bored of the book, or don't care to read? Even if you fixed this by substituting any activity you currently enjoy, you are associating a particular behaviour with a reward, which unconsciously suggests that you wouldn't perform the new habit for its own sake. This is precisely the opposite of the effect we're trying to achieve. Developing a good habit will be most successful when we do it for its own sake, when it's done automatically, and when we take satisfaction in what we've achieved – even when it's something as simple as a clean kitchen.

~~~

In theory, making habits should be easy. We do it automatically, all the time. Think back to all the examples of everyday habits we've seen earlier: travelling, eating, socializing, working, and shopping. These habits were established because we found ourselves in the same situations trying to satisfy our various needs and desires, and on returning to the same situation we made the same choice again, and so on, repetition layered on top of repetition until the habit was built. Sooner or later, the behaviour became unconscious and, whether useful or not, over time these habits took hold of sections of our lives. When making habits, we are trying
~~~

to do something similar but with conscious planning. We start with a goal in mind to associate behaviours with situations, and what we are aiming for is their unconscious and automatic execution. Each repetition takes us one small step further up the curve of that graph.

The beauty of habits is that, as they develop, they become more effortless. Even when you're tired, upset, or distracted, strong habits are likely to be performed because they're so ingrained. Habits that you've constructed yourself, for your own purposes, can seem like magic when they work. Like other behaviours that we carry out on a regular basis, the fruits of good habits may build up slowly, but they can repay the effort made to establish them many times over.

10

BREAKING HABITS

As 1985 drew to a close, like any other year, people all around the world made promises to themselves and others about how they would change in 1986. A few of those people, though, in north-eastern Pennsylvania, were prompted by a local TV news spot to dial a number and confess their resolutions to psychological researchers. Two hundred and thirteen people from all walks of life were recruited into the University of Scranton study.[1] The majority of resolutions they made to try and undo the sins of previous years were very familiar bad habits. Two-thirds of the resolutions involved weight loss (38%) and giving up smoking (30%). The next largest category, though, contained all kinds of idiosyncratic pledges: people wanted to learn to say no, to make more time for themselves, and to take more responsibility for their decisions. Participants were then contacted over the next few weeks and months to report back on whether they had kept their resolutions. The table below shows the success rate over time.

Interval	*Success*
1 Week	77%
2 Weeks	66%
3 Weeks	60%
1 Month	55%
3 Months	43%
6 Months	40%

The numbers are somewhat depressing. Almost a quarter of people had admitted failure after only a week, about half had dropped out after a month, and only 40% reported sticking to their New Year's resolutions for six months. To make matters worse, these figures are probably still optimistic because this was a self-report study, so some people probably lied about sticking to their resolutions. It's a sobering reminder of how hard habits are to change, especially when they're long established. What we need to know is why the 60% failed and what the 40% were doing right.

~~~

The strange thing about habits is that because we perform them unconsciously, we aren't always aware exactly what they are. We might well be aware of the results of bad habits,
~~~

such as being overweight, or continually missing deadlines, but how we got into the situation isn't so clear. The very first step in breaking a habit is to get a handle on when, how, and where we are performing it. Some of our own habits are obvious to us, but many are not. It's hard to change something until you know what it is in the first place. Other people may be able to provide clues (if you can bear asking), but one self-contained method is to use mindfulness.

To the uninitiated it can sound daunting, but the basic principle is simple. Being mindful is about living in the moment. In many ways, it's the exact opposite of our experience while performing a habit. Mindfulness is all about increasing your conscious awareness of what you are doing right now. It's often talked about in the context of meditation, but really it is a way of life or an attitude. Absolutely everything can be done mindfully, and paying attention is at its core. But it's not just a case of paying attention; the way in which you pay attention is also important. The attitude that's encouraged in Buddhist mindfulness techniques is affectionate, compassionate, and open-hearted. So you're not just coolly observing your own thoughts; you're also trying to be generous to them, whether they are thoughts that make you feel good or bad. You're not sitting in judgement over yourself, rather you're trying to be present and compassionate to yourself. Those who practise living in the moment say that it can give you a new way of experiencing life. Some psychologists have called it 'reperception'.[2] Reperceiving allows us to observe our automatic reactions to events and to see them clearly in context. We become more intimate with ourselves by just noticing our interpretations of the world, rather than becoming caught up in them.

Since mindfulness is such a useful mental state for habit detection, here's a quick primer on mindfulness meditation. You don't need to be meditating to spot your habits; this is about practising the right state of mind.

1. *Relax the body and the mind.* This can be done by clearing a time, sitting comfortably, possibly putting on some soothing music, and using any other (non-pharmacological) tricks for calming down that work for you. This step is relatively easy as most of us have some experience of relaxing, even if we don't get much opportunity.

2. *Concentrate on something.* Often, meditators concentrate on their breath, the feel of it going in and out, but it could be anything: your feet, a potato, a stone. The breath is handy because we carry it around with us. But whatever it is, try to focus your attention on it. When your attention wavers, and it will almost immediately, gently bring it back. Don't chide yourself, be kind to yourself. The act of concentrating on one thing is surprisingly difficult: you will feel the mental burn almost immediately. Experienced practitioners say this eases with practice.

3. *Be mindful.* This is a little cryptic, but it means

> something like this: don't pass judgement on your thoughts, let them come and go as they will (and boy will they come and go!), but try to nudge your attention back to its primary aim. It turns out that this is quite difficult because the natural tendency is to judge yourself. For example, your mind wanders back to an embarrassing moment last week and then you mentally slap yourself on the wrist. Instead, the key is to notice in a detached way what's happening, but not to get involved with it. Although this way of thinking doesn't come that naturally to most people, it has enormous benefits.

Almost everything, not just meditation, can be done mindfully. The point of trying some meditation is to practise getting into a particular type of self-observant mental state. Then you can practise brushing your teeth mindfully, surfing the Internet mindfully, even watching sport mindfully. If you can manage this every now and then throughout the day, you'll soon start to notice habits of thought and behaviour, some good and some bad. As a result, what you want to change, and why, will become more obvious.

In one study of mindfulness, participants were trying to increase the amount of vigorous exercise they did.[3] The researchers found that those who acted mindfully were more likely to act on their intentions rather than just allowing their established habits to take over. In a second group, participants were trying to cut down on their habitual binge-drinking. A very common cause of binge-drinking is social pressure: people find it difficult to stick to their intentions when other people are encouraging them to perform their habit. Once again, those who acted mindfully were better able to control their behaviour in line with their intentions.

One note of caution about practising mindfulness: it's not for everyone. Some people seem to enjoy the intellectual aspects of watching their own minds at work; others do not. If you find mindfulness tedious, then other techniques may be a better bet.

The vigilance that mindfulness encourages isn't just important in spotting how our habits operate; it's also one of the ways we spontaneously try to break bad habits. A study led by Jeffrey Quinn from Duke University looked at how effective vigilance is in breaking a habit.[4] First, they asked participants what habits they were trying to break. These included things like eating junk food, procrastinating, being late for lectures, sleeping in late, and even hair straightening. But the top three areas were related to sleeping, eating, and procrastination (need I tell you, the participants were university students). Then, they handed out diaries and asked them to record when and how they battled with their bad habits. The researchers had found that the three most commonly used strategies were vigilant monitoring (such as thinking 'don't do it'), distraction, and changing the situation.

Every couple of days, participants visited the researchers and went through each diary entry to assess the strength of the habits, the amount of temptation they felt, what strategy they had used, and whether the effort to change their habit had been successful. It emerged that only a proportion of the behaviours that participants were trying to inhibit were actually habits at all: many were temptations. The difference is important and lies in the emotions. Temptations act on our basic desires for things like water, food, and sex. When you experience temptation, say for a pie or a glass of wine, you feel it. Habits, though, while they might have initially been formed by strong emotions, are now performed unconsciously.

What the researchers found was that, for strong habits, vigilant monitoring was the most effective strategy, followed by distraction, with other approaches providing little help. This makes sense, given what we know about habits. We perform habits automatically in response to the cues from the environment, so to inhibit them we have to be on the lookout for those cues in order to exercise self-control. In a second study, the result was checked in the lab. Participants learnt one response to a word, then afterwards had to change this response in defiance of the habit they'd built up. Backing up the first study, vigilant monitoring was the most successful short-term strategy for suppressing a strong habit.

In theory, if you can spot a habit, then you can stop yourself performing it; unfortunately, there's a very hefty catch to this intuitive plan. Although this type of self-control might work in the short term, perhaps over a few days, it starts to wear down over the long term. The shortcoming is related to an irritating irony about how the mind works. This irony was demonstrated in a study, led by Daniel Wegner, which asked participants to try to avoid thinking about an imaginary white bear for five minutes.[5] Then, for the next five minutes, they were asked to think about a white bear. Throughout the experiment, participants verbalized whatever thoughts they were having and, each time they thought of a white bear, rang a bell. What they found was that participants who first tried to suppress their thoughts rang the bell almost twice as often during the second five minutes as participants in the control group, who had been thinking about a white bear for the whole ten minutes. The very act of first trying to suppress a thought made it fight back all the stronger.

Research has shown the same effects for people trying not to think about smoking, disturbing emotional memories, and for those fighting the negative thought processes in depression.[6] Wegner explains this using 'ironic processes theory'. According to this theory, here's what happens when I want to stop a recurring thought in its tracks: first, I distract myself by intentionally thinking about something else; second, and here comes the irony, my mind starts an unconscious monitoring process to check if I'm still thinking about the thing I'm not supposed to be thinking about. The trouble comes when I consciously stop trying to distract myself and the unconscious process carries on looking out for the thing I was trying to suppress. Anything it sees that looks vaguely like the target triggers the thought again, and round I go in another loop of thinking the same thought I was desperately trying to forget about.

The practical upshot for someone trying to break a bad habit is that the more you try to push the bad habit out of your mind, the more it pops up. For example, the more you suppress the bad habit of, say, eating fatty foods, then the more relevant it becomes and the more situations there seem to be in which to perform it. Not only does trying to suppress thoughts not work, but it can also have negative emotional and behavioural consequences. People suffering from a wide range of behavioural problems such as drug-taking, smoking, and overeating have been found to experience more negative emotions when they try to suppress their thoughts and feelings than if they don't.[7]

There's other evidence that trying to break habits just by suppressing thoughts may lead to exactly the opposite of the desired result. Research carried out on those with eating disorders shows that binge-eating often follows on from dieting.[8] Indeed, it's the rigid dieting (as opposed to learning the habit of eating in a restrained way) that seems to cause the binge-eating, and not the other way around.[9] Similarly, people who try to suppress the urge to drink or smoke seem more prone to a binge immediately after the suppression attempt.[10] Even regular social drinkers who are specifically trying to restrain themselves have the tendency to binge in response.[11] So sometimes inhibition can lead to habit binges rather than habit holidays. This may be why people sometimes find that when they first try to change a habit, perversely they actually start doing it more. It's handy to know that this is normal and likely to be just a phase in the process of breaking a habit.

The idea that suppressing a habit can backfire suggests an important alternative. Think of the bad habit you want to change like a river that's been following the same course for a long time. Now you want to stop it suddenly. You can't just dam the river because the water will rise up and break through. Instead, you have to encourage the river to take a different course. In order to break old habits, the attempt needs to be paired with making a new habit.[12] This is why people trying to give up smoking chew gum: it's not just the nicotine that's probably in the gum; it's also replacing the habit of having something in the mouth. Or, take another simple example: say you've got a habit of avoiding the cracks in the pavement that you want to change. Just inhibiting this habit will be difficult, because the more you try to avoid thinking about the cracks the more you'll notice them and be tempted to perform your habit. Instead, you could decide to start a new habit of looking upwards. If you can inhibit your old habit for a while, this creates a window of change into which you can insert a new, desirable habit. With repeated performance, this new habit will slowly take over from the old one.

When you choose a new habit to help break an old one, you can use the ideas already discussed. Mental contrasting can help you think about which habit will successfully replace the old one and implementation intentions, or 'if-then plans', can be used to make very specific plans of action that link old situations to new behaviours. You can also use implementation intentions to help shield the habit from intrusive thoughts or slip-ups, and to think about how you'll address dissatisfaction.

Even with a new habit established, the old one will remain there, lurking in the background. Old habits really do die hard. Research on both animals and humans suggests that even after habits have apparently faded away through lack of repetition, they still lie in wait to be reactivated.[13] It's like the fact that you never forget how to ride a bike: those connections in your brain aren't gone, they're just dormant. Unfortunately, making a new habit doesn't generally destroy the old habit. Highly familiar contexts will still have the power to cue up old patterns of behaviour while even roughly similar contexts can be dangerous. Habits which fulfil strong needs or solve problems in the short term are particularly prone to relapse. That's why, for example, alcoholics have to be continuously vigilant, as their drinking can easily be cued by low mood or walking past an off-licence.

Highly practised habits resist the will simply by being unconscious. We perform them automatically without realizing. There is evidence, though, that implementation intentions help bring the choice about whether to perform a habit up into the conscious mind.[14] Making very specific plans in advance reduces the amount of thinking that must be done in the moment and provides an automatic response to compete with the bad habit.[15, 16] For very strong habits, even finding that you have a conscious choice is a step up from just performing your habit automatically.

Fortunately, most everyday habits have all kinds of alternatives: eating fruit rather than ice cream, or reading an improving book instead of watching TV, or asking your partner unusual questions over dinner rather than running through the same old subjects. Successfully breaking a habit is much more likely when you have a shiny new well-planned habit to focus on rather than just thinking about suppressing the old habit.

~~~

No one likes to think they're average, least of all below average. When asked by psychologists, most people rate themselves above average on all manner of measures including intelligence, looks, health, and so on. Self-control is no different: people consistently overestimate their ability to control themselves.[17] This overconfidence can lead people to assume they'll be able to control themselves in situations in which, it turns out, they can't. This is why trying to stop an unwanted habit can be an extremely frustrating task. Over the days and weeks from our resolution to signs of change, we start to notice it popping up again and again. The old habit's well-practised performance is beating our conscious desire for change into submission.

People naturally vary in the amount of self-control they have, so some will find it more difficult than others to break a habit. But everyone's self-control is a limited resource. It's like muscle strength: the more we use it, the less remains in the tank, until we replenish it with rest. In one study of self-control, participants first had to resist the temptation to eat chocolate (they had a radish instead); then they were given a frustrating task to do. The test was to see how long they would persist.[18] Radish-eaters only persisted on the task for about 8 minutes, while those who had gorged on chocolate kept going for 19 minutes. The mere act of exerting will-
~~~

power saps the strength for future attempts. These sorts of findings have been repeated again and again using different circumstances.[19]

We face these sorts of willpower-depleting events all day long. When someone jostles you in the street and you resist the urge to shout at them, or when you feel exhausted at work but push on with your email: these all take their toll. The worse your day, the more the willpower muscle is exerted, the more we rely on autopilot, which means increased performance of habits. It's crucial to respect the fact that self-control is a limited resource and that you are likely to overestimate its strength. Recognizing when your levels of self-control are low means you can make specific plans for those times.

The good news is that, although willpower is a limited resource, there are all sorts of strategies you can use to help in breaking an old habit. Pre-commitment is one tool for winning the battle for self-control. Say you want to avoid your bad habit of wasting a weekend playing computer games. An excellent pre-commitment strategy would be to take your games console out into the garden and smash it to pieces. That's an extreme measure which represents serious commitment; an alternative might be leaving the console at a friend's house. This is a way of restricting the choices of your future self. By making the hard decision when your self-control is high, you can protect yourself against a later time when your self-control has taken a battering at work or from the commute. Even far less drastic measures like avoiding the procrastination habit by simply setting yourself deadlines can be effective in helping self-control.[20] The power of pre-commitment has also been shown to increase people's money-saving habits.[21]

Self-imposed rewards and penalties can also work, with the proviso that doing something for its own sake is much better than relying on the carrot and stick.[22] Even the most basic and much-repeated of all self-help tips can be useful: Yes, think positive! If you can generate more optimistic predictions about your ability to change your habits, then it will boost your motivation.[23] Merely giving the instruction to 'think positive' is a bit vague, so let's break that down. One way of generating a more optimistic outlook is to think about your attitudes towards the habit you want to change. What is it about your goal of breaking and changing a habit that really attracts you? At the same time, what is it about the bad habit that most repels you? The more relevant and vivid you can make both the positive aspects of your new good habit and the negative aspects of the bad habit, the more likely you are to exercise self-control.[24] To give a concrete example: people who are deeply disgusted by their own nail-biting, and would love to have long nails, have a much better chance of breaking the habit than those who think it's no big deal.

Despite all these intersecting plans and strategies, even the most strong-willed of us will suddenly find ourselves performing that bad habit we promised ourselves was gone for ever. Psychologists have found, though, that self-affirmation – thinking about your positive traits – can help in the quest for control of the self. In an experiment by Brandon Schmeichel and Kathleen Vohs, half of their participants had their self-control depleted by writing a story without using the letters 'a' and 'n'. They were all then given a classic test of self-control: submerging their hands for as long as they could in a bucket of icy cold water, which becomes very painful after a minute or two.[25] Those people who had written the story without a's and n's, and had no chance to recover, were only able to keep their hands in the water for an average of 27 seconds, compared with 80 seconds for those who'd been able to use any letters. Here's where the replenishing effect of self-affirmation comes in. Before putting their hands in the freezing water, half the group with depleted self-control wrote about their core values, such as their relationship with their family, their creativity, or their aesthetic preferences, whatever they felt was important to them. After this, they managed to hold their hands underwater for an average of 61 seconds, more than doubling the average time of the unreplenished group. So it seems that self-affirmation can refuel depleted self-control.

The reason this works is because of how it changes our mind-set. Concentrating on core values tends to make people think more abstractly. When we are philosophical about what we're doing, it helps distance us from the temptations of the moment, allowing our self-control a chance to intervene in bad habits. The self-affirmation trick works both in the moment as well as when planning for the future, since it encourages people to find ways to protect their goals from temptations they know they will face.[26]

All of these efforts to bolster your self-control have a

happy side benefit. Just like a muscle, exercising self-control increases its power. So merely trying to change a habit will have a snowball effect. A study on people successfully following a new exercise programme showed that their self-control also increased, but crucially to areas that had nothing to do with exercising.[27] After practising their self-control, they were more likely to choose studying over television and developed better domestic habits, like washing the dishes more frequently. The same was true in another study of people who were being taught money management.[28] Participants found that not only did they get more control of their spending habits, but they also found it easier to regulate their alcoholic intake, their emotions, their eating, and, once again, their domestic habits. Clearly, working your self-control muscle can be useful in all sorts of ways.

~~~

Making plans and exercising your self-control may only get you so far in breaking old habits: more drastic measures could be required. One of the best ways of changing an existing habit is to change the situation. Since habits are cued up by the situations we routinely find ourselves in, then changing that situation should avoid the habit cue. It's like when you go on holiday to a new, unfamiliar city and suddenly it's both exhilarating and exhausting. At home, eating, trips out, and even conversations are partially or completely automatic, but on holiday, without the crutch of familiar situations, even the smallest decisions have to be made consciously. Soon enough, you build up habits: there's one par-
~~~

ticular café where you enjoy the coffee and the view; you start arriving at the beach at the same time and returning to the same restaurant in the evening. Still, for a brief moment at the start of the holiday there is a window of opportunity, when you're all at sea in a new situation, where anything could happen and you can leave your old habits behind.

There's certainly evidence that a change of context can help change habits. In the earlier quoted study of students who moved from one university to another, they were able to change their everyday TV watching, reading, and exercising habits with their change of context.[29] The students' intentions had also changed between the two locations, and this was not a coincidence. New situations force us to think consciously about what to do, and so our actions are tightly coupled to our intentions. That's also why people's shopping habits change with major shifts in their lives such as moving house, switching jobs, or having a baby. The same is true of travel habits. A British study has looked at how moving house affected travel choices.[30] They found that people who had moved house recently were more likely to change their travel habits when motivated by environmental concern. In comparison, those who had the same high level of environmental concern but had not moved house recently were less likely to make a change. Once again, something about being in new surroundings is enough to allow us to jump out of old ruts and start creating new ones.

It's a bit drastic, though: moving to another house, city, or country just so you can walk to work rather than drive. And while holidays might be good for new habits, we all come home soon enough, back to the old environment and the old habits. The vast majority of us don't have the desire or, indeed, the resources to relocate just to shake up old habits. Still, there are ways to make more subtle adjustments to our existing environments without these kinds of upheavals.

A couple of clues come from a very simple set of studies with real potential for personal change. Public-health officials have been obsessed with getting people to use the stairs for decades. It's seen as the kind of exercise that can be easily incorporated into day-to-day life. All sorts of tricks have been tried, but one of the simplest is also the most effective. A sign is put at the bottom of the stairs telling people that walking up the stairs burns about five times as many calories as taking the lift. And magically, for the price of a piece of paper, some people do switch from the lift to the stairs. When sixteen studies involving this intervention were analysed, they found that, on average, stair use increased by 50%.[31] Admittedly, this is from a pretty low baseline, because not many people generally use the stairs in the first place; but it does demonstrate that this small change to the environment can work.

People sometimes spontaneously use these types of environmental interventions. Alarm clocks are moved out of arms' reach, fatty foods are removed from the house, or the Internet is unplugged and the router hidden so that some real work can be done. All these types of small changes to the environment can help remind us of the established habits we are trying to break. The key is to find a way to sabotage our unconscious, automatic processes and bring the decision up into the conscious mind. When there are no crisps in the house, no Internet connection, and you can't hit the snooze button easily, it forces you to recall your promise to yourself and consciously decide whether you want to break that habit. You might still make the wrong choice, but at least you'll have a conscious choice rather than just performing the bad habit automatically.

The problem with small changes to the environment, like notes and even alarm clocks, is that they quickly lose their novelty and become easy to ignore. Anyone who has ever lived in a shared house knows that notes left lying around the place are soon overlooked. If you notice that you're failing to perform your new habit successfully, have a look at the reminders you've set up in your environment. Is it possible that you've started ignoring them? If so, it's time to change the reminder to something you will notice. Ultimately, though, whether or not these sorts of environmental changes are taken seriously depends on the level of commitment to establishing a new habit in the first place.[32] Notes and other environmental tweaks can jog the memory but they can't force us to perform the new habit.

~~~

There's little doubt that even the simplest habit changes can tie us up in knots. Over a century ago, the German-American psychologist Hugo Münsterberg experimented on himself tirelessly in the quest to break personal habits. He tried making simple changes to his daily routines and meticulously recorded each time he reverted to his old ways.[33] In one
~~~

attempt, he decided to stop using the door from his office leading on to the corridor and start going in and out through a different door to his secretary's office. Before long, and to his intense annoyance, he caught himself in the process of using the hallway door, not just once but several times.

On another habit-changing spree, Münsterberg began experimenting with the position of his inkwell. He was accustomed to dipping his quill in a well to the left of him, so he placed an empty well there and put a full well on the right. Over a full day's writing, he noted each time he reverted to the old habit. After about a week, he'd incorrectly gone to the left 64 times, but after two further weeks, this came down dramatically. Subsequently, he practised swapping from left to right until, eventually, he was able to alternate the full inkwell from left to right without making any false moves. He repeated the same process with his pocket-watch and was gratified to see that with persistence he could switch between habits without making any mistakes.

Although many people's attempts to break their bad habits fail, there is hope. We all manage to change them from time to time: research into everyday habits like eating breakfast, vegetable consumption, and travel choices has shown that we can, and do, manage to break our old habits, or at least replace them with good new ones.

Like making habits, breaking habits requires a bit of cunning, indeed more so, because the old habit is always lying in wait, ready to be reactivated. With an understanding of how habits operate, though, you have a much better chance of making the change stick.

11

HEALTHY HABITS

Are you feeling fat? Are you depressed about how you look? Does your weight yo-yo as you go on one crash diet after another? Are you fed up with feeling hungry? Do you want to know the solution to all your weight-loss problems? Well, let me tell you a secret! It all began one day fifteen years ago when a friend asked me for advice about losing weight. I told her that the secret of my toned figure was down to a special diet I have designed. She was amazed as I explained its completely natural components and that she could lose all her excess weight in only twenty-eight days! She could hardly believe it, but four weeks later she was wearing a new dress three sizes smaller! Seeing how excited she was, I decided to let everyone in on my diet secrets. Now

I've helped millions of people all around the world change their lives. And I can help you, too...

Of course, I'm taking a cheap shot at the worst type of diet book, the ones that make the biggest promises with the weirdest gimmicks, while suggesting that the whole process will be a breeze. Crash diets may work in the short term, but we all know in our heart of hearts that the celery-and-gravel-type diets will never stick in the long run. Putting the more ridiculous end of the diet-book spectrum aside, though, there is plenty of very good advice out there. Many books about healthy eating don't try to sell you a miracle cure based on eating or not eating some particular foodstuff. Instead, they offer much sensible advice about what to eat and the other components of a healthy lifestyle. And people continue to buy good diet books, watch TV programmes about healthy eating, and surf the Internet for healthy recipes; it's not as if the information about maintaining a healthy lifestyle is hard to find. Still, many people seem unable to take this good advice.

By now, you'll have heard all the obesity scare stories. We all know how much overeating is costing in terms of treating the resulting diseases, and how little people are managing to change their eating habits. Here's just one fact from many: since 1980, the percentage of US adults who are obese has doubled, while the percentage who are severely obese has quadrupled.[1] These sorts of figures are mirrored in many other countries around the world.[2] Governments have tried to change our ways by targeting us with healthy-eating campaigns, but these have a fatal flaw. That flaw is simply that they try to influence us by changing our intentions. They warn us about the dangers of obesity, what it's doing to our health, and why we should change. The trouble is that almost everyone knows what they are doing to their health and why they should change, but they don't seem to be able to manage it. According to some estimates, only 20% of people stick to weight-loss diets in the long term.[3]

It's not that public health campaigns are a waste of time: in some areas, they can be modestly effective. The problem is that when it comes to strong habits, they only have a marginal effect.[4] Given what we now know about how habits operate, this makes perfect sense. We know that people buy mostly the same foods each week, eat that food in the same environment, and consequently, their eating is mostly controlled by habits.[5] We also know that attempting to educate people – that is, change their eating intentions – only has small effects on their behaviour in the face of strong habits.[6] For example, participants in one study who were in the habit of visiting fast-food restaurants found them very difficult to avoid, even when they tried.[7] The only situation in which it's easy to change your eating habits is when no pattern has been established. So, good news for someone who was born yesterday.

The rest of us have to compete with the hard taskmaster of our established eating habits. The sheer power of habits is beautifully demonstrated in one study on popcorn.[8] This focused on how some people eat popcorn at the cinema: typically in massive quantities and with no regard for how hungry they are. What the researchers wanted to see was how the context and their established habits affected popcorn consumption. To do this, they had some participants sit in a cinema watching trailers while others sat in a meeting room watching music videos. None of them was aware that the study was about eating habits; they were told it was about attitudes and personality.

When sitting in the cinema, strong habits cued by familiar circumstances had their familiar effect: people behaved like popcorn-eating robots. In the cinema, it didn't matter whether the popcorn was stale or fresh or whether people were starving or full, they still munched about the same amount of popcorn. Habit even steamrollered preferences: liking for popcorn had very little effect on how much they ate. When popcorn-eating habits were weaker, though, people behaved more like rational, thoughtful human beings by eating less of the stale popcorn.

In contrast, the participants in the meeting room *all* behaved more like rational, thoughtful human beings, whether or not they had a strong habit of eating popcorn at the cinema. They ate less stale popcorn, and less overall if they weren't hungry. Even for those people who had strong popcorn habits, the change of situation was enough to disrupt their automatic behaviour. Overall, people in the meeting room ate 50% less popcorn compared with those in the cinema.

So can the psychology of habits help us change our eating habits? Obviously, it can't solve the obesity crisis in one go, but it can provide some very useful and practical insights. One of the biggest mistakes people make, and one that's often repeated by bad diet books, is trying to make drastic changes. Sustaining crash diets or massive shifts in what we do repeatedly is likely to be much too difficult for the vast majority of people. We've already seen how hard it is to make or break one new habit; imagine if you had to change many of your eating habits all in one go. It's just not practical; they're never going to stick.

Evidence for how more modest habit shifts can be effective comes from the US National Weight Control Registry, which since 1995 has been tracking the weight-loss attempts of thousands of Americans.[9] Like other studies, it finds that only a small percentage of people are able to lose weight and keep it off. Amongst those people who do manage it, though, one key factor is establishing regular, unchanging routines. Successful dieters regularly ate breakfast, ate the same over the weekend as they did during the week, and ate the same types of foods, mostly in the same environment (at home).[10] Crucially, many dieters weren't able to eat all the right foods all the time and meet all the goals they set for themselves, but those who were successful in the long term at least made some efforts in the right direction and managed to maintain these efforts. They changed one small habit, say, eating an apple in the morning, established that, then moved on to other incremental changes. This isn't exactly the stuff of Hollywood films, but it does have the advantage of being true, and actually working.

On top of concentrating on small changes, we've seen how habits are automatically activated by our environments. We know that people make dozens and dozens of choices about food each day, and we can't hope to stop and make a conscious, informed decision every time. This is especially true when you get home hungry, tired, and still thinking about work – then your habits will be in full control. But with this knowledge it's possible to tweak the environment to encourage the right kinds of healthy habits.

Go into your kitchen and have a look around: what do you see? Are the first foods that greet your eye healthy or unhealthy? Do you have a bowl of fresh fruit on the counter, or snack foods? How big are the plates you eat off? How much food do you have stockpiled in the cupboards? Are the containers they are stored in big or small? Research has found that when food is stored in bigger containers, people tend to eat more of it because of 'stock pressure' (the costs of storage).[11] People also tend to eat more of food that is visible and close to them.[12] They eat more from larger plates without noticing, spontaneously giving themselves much larger portions and eating considerably more of the food they've served themselves.[13] The same is true of drinks, with people drinking more from bigger containers.[14] And so the list goes on... people eat more when the packaging is larger, the utensils are bigger, and they have bigger serving spoons.[15] This suggests that there are a whole series of environmental changes that can encourage healthy eating. They'll need experimenting with to make them right for you, but they should be relatively easy to make.

Even outside the home it's possible to use little tweaks to adjust your habitual behaviour. Here's a little trick suggested in a follow-up to the popcorn-related study. Some people in the cinema were told to eat with their non-dominant hand, so if they were right-handed, to eat with their left. Magically, this seemed to have the desired effect of jolting people out of their habitual behaviour and bringing the conscious mind back into action. By eating with their non-dominant hand, participants once again responded to the freshness of the popcorn they'd been ignoring while eating in a trance with their dominant hand.

Whatever changes you make to your own environment, whether inside or outside the home, huge numbers of cues for bad eating habits will always be left: shops, adverts, vending machines, and so on are everywhere. That means that changing eating habits will always be about self-control to a certain extent. Self-control can be built up by managing to stick to small changes in eating habits, and so benefits should lead to further habit change. Studies suggest that self-control is a generalized ability: the practice of self-control doesn't have to be related to food, but the benefits will still seep through. That along with the other psychological self-control techniques discussed previously are all likely to be helpful: respecting the fact that self-control is a limited resource, pre-commitment, rewards and penalties and using self-affirmation in moments of weakness. In particular, practising mindfulness can be helpful in noticing temptations and allowing them to pass without taking any action.[16]

Once again, implementation intentions – making highly specific plans – can be useful in changing eating habits. All the same guidelines to forming implementation intentions discussed in the previous two chapters apply to eating, but with one big proviso. One common type of plan people make when dieting is a negative one. They say to themselves something like: I must not eat chocolate. This is a kind of implementation intention, but is it a good one? This has been tested in a series of studies which found that when people used this kind of negative implementation intention, it had an ironic rebound effect.[17] Recall from the last chapter that trying not to think about a white bear tends to make the thought come back stronger. The same thing is true when participants tried not to think about chocolate. Instead of putting the thought out of their head, the reverse happened: when a snacking situation arose, they thought about chocolate even more. Worse, though, this rebound effect was strongest when participants' snacking habits were also strong.

Participants in this study were better able to avoid thoughts about unhealthy snacking if they avoided negative implementation intentions. Instead, it was generally more effective to make a positive implementation to do something else. Instead of thinking: '*If* I'm hungry between meals, *then* I must avoid chocolate', it's better to think: '*If* I'm hungry between meals, *then* I will eat an apple'.

Even when formulated correctly, though, implementation intentions are still often too weak to have an effect on existing habits. The problem is that so much of our eating behaviour is automatic. The decision about what to eat sometimes isn't even available to our conscious mind. Worse, we tend not to notice what cues habitual, unhealthy eating behaviours. So psychologists have come up with a stronger combination of strategies to try to increase awareness of the cues. This involves using implementation intentions along with the previously described strategy of mental contrasting: imagining how changing your habit would be beneficial, then contrasting this with the negative reality.[18] The idea is to make the association stronger in the mind between the bad eating habit and how this is stopping you from reaching a cherished goal.

This combination strategy has been tested in a study on snacking habits. Here are the instructions from the study, as they provide a useful exercise:

> Sometimes a wish does not become fully realized, even if one is very motivated to realize the wish. What situations could make it hard for you to diminish your bad snacking habit? Think about which is the most important obstacle to eating fewer unhealthy snacks for you personally and write it down in one keyword. Now, depict in your thoughts the events and experiences that you associate with this obstacle. Give your thoughts and imagination full scope and write them down.[19]

After carrying out this exercise, participants set their implementation intentions while avoiding negative constructions so as to sidestep the ironic rebound effect. Here's an example: 'If I [obstacle] and I feel like having a snack, then I will eat a(n) [choice of fruit]'. Typical obstacles could be thoughts or emotions like being bored, hungry, or tired;

or could be situations, like passing a fast-food outlet. The results of this study suggested it can help people avoid unhealthy snacking. When participants' habits were strong and they tried the mental contrasting along with the replacement implementation intention, they consumed fewer unhealthy snacks and almost half as many calories. The reason this technique works is that it involves identifying the exact situations that cue snacking behaviour. Without identifying the cue, a strong habit is likely to be performed whether or not implementation intentions are made.

Since eating habits are so well established, though, they are ready to escape from the cupboard we've tried to lock them in at any moment. Changing them should be a gradual process, carried out over a period of time. The psychological methods discussed here involve trying to identify the situations that spark your eating habits, as well as bringing some of your automatic eating behaviours up into the conscious mind. If old habits start to reassert themselves, then it's time to make small changes to your implementation intentions, or the situations in which you perform your habits. It's about finding what works for you to help break old habits and start building healthier eating habits.

~~~

As a society, we have a love/hate relationship with exercise: well-meaning people and organizations love to give us advice and we hate to take it. Just as with what we eat, engaging in regular exercise has the potential to improve our lives dramatically. Even if you ignore the improvements to physical
~~~

health, exercise is worth it for the psychological benefits alone. Dozens of studies involving thousands of participants have shown that exercise can improve memory, attention, reasoning, planning, and the overall speed at which your mind works.[20] True, it might not turn you into a genius in a couple of weeks, but exercise is probably better than the other methods for increasing cognitive function that are around, like computer-based 'brain training', drugs, and nutritional supplements. On top of this, exercise makes you feel good which is why it is often 'prescribed' by doctors as one way of helping with depression, anxiety, or eating disorders.[21] It may even be as effective as antidepressant medication or having cognitive therapy.[22] We know it, everyone tells us, and yet people find it hard to establish regular exercise habits.

For many years psychologists thought that our exercise habits were mostly or completely under our conscious control. They thought that the reason we don't exercise is because we don't want to. This view is changing because, as we've seen across many different areas of our lives, we may intend to start exercising, but that doesn't always translate into the actual behaviour. Some health psychologists think that the sorts of interventions which only rely on changing your attitudes to exercise can't hope to break down strong habits, like driving to work every day when you could walk or cycle.[23] Instead, exercise has a large habitual component, and exercise habits (or rather, laziness habits) are established when we're young. Indeed, the simple question of whether we exercised when young may have one of the largest influences on whether we exercise later in life.[24]

Research into specifically how habits can be changed in exercise is at an early stage. However, we do know what types of interventions psychologists have found to be effective in helping people exercise more, and the results have a familiar ring. Research led by Susan Mitchie at University College London has pulled together over 100 different studies on both healthy eating and exercise that involved almost 50,000 participants.[25] All kinds of techniques were tested to try to get people eating healthily and exercising more, including simply providing encouragement, to training in time management, and providing warnings about the dangers of unhealthy behaviours. Out of the 26 different techniques that were studied, though, one emerged head and shoulders above the rest: it was self-monitoring.

Take one Scottish study, led by Graham Baker, that aimed to encourage more walking.[26] Participants were first asked to think about why walking would be beneficial for them, then what might stop them from reaching their goal (notice that this is very similar to the mental-contrasting procedure discussed earlier). Then they made specific plans about when they were going to exercise, and set themselves goals to reach. Crucially, though, they were also given pedometers: devices that measure the number of steps taken. This allowed people to see exactly how much exercise they were engaging in. We can't tell specifically from this study that it was the pedometers that made a difference, but it was probably an important component. Anything that enhances awareness of our exercise habits is likely to be helpful, whether it's developing a new habit or trying to break and replace an old one.

As well as noticing your own behaviour, we've seen that making and breaking habits requires (the dreaded) self-control. Mitchie's research found that, for exercise, some of the techniques we've already explored for bolstering self-control were helpful. Using implementation intentions to decide when and how to exercise is one of the most important. This is because, if exercise behaviours are repeated in similar circumstances, then they are more likely to become habitual. If you always run after work and before supper, or always visit the gym in the morning before breakfast, then, with repetition, these new habits are much more likely to stick.

Implementation intentions can also be used to guard against any thoughts or circumstances that will put you off exercising. These types of simple plans have been shown to work and encourage more exercising.[27] For example, many people say they are too busy to exercise. The implementation intention you can use for this is: '*If* I put off exercising because I'm too busy, *then* I'll remind myself that exercising clears my mind, allowing me to work more efficiently'. Another unbelievably common mistake is to set your goals too high, too soon (the exercise equivalent of a crash diet). Here's the implementation intention you could use for that: '*If* I'm tempted to push myself too hard initially, *then* I will try and set realistic goals'. Other problems you might face are that it's too cold outside, you don't have the right equipment, you don't want to exercise on your own, or you're anxious or depressed. All of these have solutions and, with a little creativity, you can work out which are best for you.

One major obstacle to launching the exercise habit is that

it's difficult to incorporate into the day. With eating habits, we have a natural reminder to eat: our appetites. These come round regularly every four or five hours. For people who aren't in the habit of exercising, though, there's no physical desire that needs satiating, and so there isn't a regular cue that we should perform a particular behaviour. Developing the exercise habit is about creating that cue or alert that automatically makes us exercise. Finding the right slot in your day, the right type of exercise, and dealing with the inevitable barriers will determine whether or not you can make the habit stick. Exercise habits may take months to ingrain, but the improvements to your life can be considerable.

~~~

If you thought changing exercise or eating habits was difficult, then check out smoking. Some statistics from the UK – which is at the forefront of smoking reduction – tell a sobering story. In England, about one-fifth of people over 16 smoke, making 8.5 million smokers.[28] Of those, 3.9 million tried to stop smoking in 2009 – typically, only 2 to 3% succeed; that's just 136,000 people.[29] Within one week of trying to give up, 75% of people have started smoking again. Within one month, 90% are back to puffing away. Based on these numbers, the success rate is even worse than for those dieting.

Just as none of us needs telling about healthy diets or getting some exercise, none of us needs reminding about the dangers of smoking. There's the cancer, the heart disease, the emphysema, and all the rest. Here's just one fact that can
~~~

be a useful motivator for more mature smokers: for every year that you delay giving up after the age of 35, you lose, on average, three months of life expectancy.

Given that smoking is so bad for you, and there can't be anyone reading this book who doesn't know it, why do people find it so hard to give up? The answer is that smokers trying to quit are battling two cast-iron habits at the same time. There's the type of habit that's related to automatic, habitual responses to situations (like smoking with friends or while drinking), and then there's the chemical habit. It's the chemical habit, or addiction, that makes it so hard to quit. When you smoke, a bolus dose of nicotine hits the brain within seconds, which releases the neurotransmitter dopamine in the midbrain.[30] With continued smoking, the brain's chemistry changes to create a hunger for nicotine. When a smoker tries to quit, they aren't just battling a behaviour associated with a particular situation; they are also fighting their own chemistry. Smokers who quit will experience a whole range of unpleasant symptoms within the first week: anxiety, anger, insomnia, impatience, restlessness, and depression.[31] All of these can be relieved by having a cigarette. It's no wonder that 90% of people quit quitting within the first week.

People who are quitting usually try to beat the chemical habit with nicotine replacement therapy (NRT). Whether it's a patch, gum, inhaler, or spray, when nicotine is delivered by some other method than cigarettes, it helps people to quit. Indeed, nicotine replacement therapy can increase the chances of quitting successfully by between 50 to 70%.[32]

Using both the patches and the gum together increases the chances again.

NRT may help with the chemical habit, but it makes little difference to the behavioural habit. Strong habits are still powerful enough to cue behaviour automatically, even when it's illegal. In the UK, a total smoking ban in public places came into force in July 2007, which meant that smokers visiting pubs now had to take a trip outdoors to light up. Despite the ban, researchers wondered whether the habit would be so strong that smokers would still light up indoors by accident.[33] To test this, they measured the smoking habits of 583 people; then, after the smoking ban, asked them how many times they had accidentally lit a cigarette or nearly done so while inside a pub. Almost half admitted they had, and those with a stronger smoking habit were more likely to find themselves lighting up automatically before remembering that it was illegal. Importantly, this was probably unrelated to nicotine dependence as heavier smokers were no more likely to make the mistake than light smokers. Instead, it was related to alcohol consumption: it was the strong association between drinking and smoking that led people to light up automatically.

The strength of smoking habits means that prospective quitters have to think about how they are going to deal with automatic behaviours. Many, many different types of techniques have been tried to help people stop smoking. Broadly, though, beating the behavioural bad habit means boosting motivation and supporting self-control.[34] Without motivation, of course, no attempt to change is going to succeed.

Much motivation to change can be drawn from the well-known dangers of smoking and its obvious disadvantages such as the smell and the yellow teeth. But these need to be balanced against the barriers to change – these are likely to be personal and environmental cues which set off the behaviour.

We've already seen how mental contrasting can increase motivation. This has been tested in smokers by asking them to think about four benefits of giving up smoking: things like having better skin and more energy and self-respect.[35] Then, they took one of these and thought about all the aspects of their reality that would stop them from giving up. These were things like peer pressure, stress, and partying. The results showed that those who contrasted fantasy with reality, in contrast to those who didn't, were more likely to take immediate action in trying to quit. Crucially, this only works for people with high expectations of success. When expectations of success are low, the positive fantasy looks too far away from the negative reality. This is why people often need extra support when trying to quit smoking. Studies have shown that getting behavioural support can double the chances of successfully quitting over and above NRT.[36] In addition, individual counselling and telephone support can also be effective.

In the self-control category are many of the techniques that will be familiar from healthy eating and exercise: making specific implementation plans about when and how to quit, where to get NRT medication, and how to take it. In particular, smokers are encouraged to think about the places, people, and routines that are cuing their smoking habit and what is likely to cause a relapse. As with other bad habits, it can be difficult to break the smoking habit without replacing it with another one.

This careful planning process can be seen in action in one study on the effects of implementation intentions for smokers.[37] Prospective quitters were given a list of twenty situations in which people are often tempted to smoke. For example, when getting up in the morning, in an emotional crisis, and so on. Then there were a list of twenty potential ways of acting. The key is to link up the dangerous situations with alternatives to smoking. For example, '*If* I am tempted to have a cigarette when I wake up, *then* I will remember that I get upset when I think about my smoking'. The results of this study showed that participants using the implementation intentions were more likely to quit than a control group, and were also more likely to cut down their smoking of cigarettes. Similarly, implementation intentions have also been successfully used to help stop adolescents from starting to smoke in the first place.[38] Overall, focusing on ways of changing routines has been found by the NHS to be amongst the most useful approaches to quitting smoking.[39]

One factor that may well be helping people to quit is the bans on smoking in public places that have been introduced in many countries. Although these were primarily designed to protect non-smokers from the dangers of passive smoking, they have become a good test of how environmental changes can affect habits. If smoking is a habit cued by the environment – like being in a pub drinking with friends or having just finished a meal in a restaurant – then making it illegal should curb the habit. In the UK, there is evidence that the decline in smoking was greater in 2007–2008 than in any previous year.[40] UK researchers have also been into pubs before and after the ban to count the number of people who were smoking and how many went on to give up after the ban.[41] They found that 15% of people in their study had quit, which is a very high rate given that the average is around 2 to 3%. However, the figure of 15% has to take into account the fact that people often relapse. Nevertheless, research in other countries has also found this beneficial effect for workplace smoking bans.[42] Research in Germany found that the stricter the bans were – in other words, the more places in which people were prohibited from smoking – the greater the decreases in smoking.[43]

~~~~~~~~~

Whether it's trying to eat more healthily, getting some exercise, or quitting smoking, it's obvious how hard it can be to develop healthy habits. The small percentage of people who successfully manage to make the change clearly tells the story. Part of the reason, though, that so many people don't manage to change is that they're not aware exactly how the power of habit controls their behaviour. It's easy to overlook environmental cues, and people are naturally reluctant to admit that they don't have complete control over themselves. What the psychological research into healthy habits reveals, though, is that understanding and working with the unconscious, automatic nature of habits can help people to make a change.
~~~~~~~~~

Making healthy habits should be a voyage of discovery. It's both about discovering exactly what you're already doing, then working out what sorts of modest changes are practical. Changing eating habits is about noticing what you eat, when and why, and making small adjustments that are sustainable. Changing exercising habits is about finding a regular slot in your day and dealing with the inevitable mental barriers you'll face. Quitting smoking is about more than chemical addiction; it's also a forensic analysis of what causes you to light up and what you're going to do instead. Change requires commitment, which is why the mental-contrasting exercise can be so useful in clarifying your goals.

The true aim of personal change is to turn our minds away from miracle cures and quick fixes, and to adopt a long-term strategy. Habit change isn't a sprint; it's a marathon. The right mind-set is to wake up tomorrow almost exactly the same person, except for one small change – a small change that you can replicate every day until you don't notice it any more, at which point it's time to plan another small change...

12

CREATIVE HABITS

Creativity is mysterious. Just ask any scientist, artist, writer, or other highly creative person to explain how they come up with brilliant ideas, and, if they're honest, they don't really know. Having got used to these sorts of questions over the years, some will provide well-rehearsed answers which probably have less to do with the truth and more to do with fulfilling the audience's expectation of an artistic-sounding answer.

In 1952, the American poet and academic Brewster Ghiselin published his book *The Creative Process*.[1] In it, he asked how brilliant people – Albert Einstein, Vincent Van Gogh, Pablo Picasso – managed to be so creative. He concluded that it was extremely rare for anyone to force themselves consciously to create a new idea. Instead, the most innovative ideas seemed to bubble up from the unconscious without any purely conscious, calculated process. The poet William Blake, for example, claimed to be describing images he saw in his mind, while Mozart said he wrote down melodies that just came to him. We get the same frustrating answers from contemporary writers, artists, and other creative people: they don't seem to know where their ideas come from.

Psychologists have approached the question of how to be creative in a different way. Instead of asking people directly, they have examined the psychological conditions of creativity. Psychologists ask under what circumstances people are at their most creative. They tweak the circumstances and measure the results so that over hundreds of studies carried out across decades, what has emerged is a picture of which psychological circumstances are most associated with heightened creativity. It turns out that these creative circumstances have an important connection with habits.

~~~

A simple creativity test invented more than 80 years ago gives us an insight into how habits and creativity interact. It's called the two-string problem.[2] Participants are led into a room with two strings hanging from the ceiling. One of the strings is hung in the centre of the room and the other near the wall. People are told that their goal is to try to tie the two strings together. They soon find the problem. Even with one string in hand, they can't reach the other string
~~~

without dropping the first one. To help reach their goal, participants are allowed to use any of the objects lying around in the room. Amongst these are tables, poles, chairs, clamps, and a pair of pliers. One of the solutions participants found involves tying the pliers to one of the strings, then making it swing from side to side. They can then grab the other string and wait for the pair of pliers to swing back towards them. It's a neat solution, but it's hard to arrive at because of what we already know. Pliers are for gripping and possibly cutting, not for using as a pendulum. Solving the problem creatively involves seeing past the habitual functions of objects and ideas. Realizing that the pliers could be used as a pendulum is a creative solution that's inhibited by habit.

This puzzle is similar to real-world problems in two ways. First, participants don't have quite the right tools for the job. That's what life is like. You have to adapt whatever is there to the problem at hand. Second, there is more than one right answer. These are two crucial features of creative problem-solving: adapting objects or ideas to new uses, and the fact that problems can be solved in many ways.

Let's take a modern, Internet-age example of how habit can inhibit creativity. Once upon a time, not so long ago, people thought the Internet was like a big, virtual encyclopedia. When you want to find things in an encyclopedia, you can look them up in the index. So, quite naturally, some reasoned that what the web needed was an index, a list of its contents broken down by category. Huge amounts of time and money were invested in building these indexes on the assumption that people would find what they wanted by drilling down through hierarchies of information.

Today, indexes are still used, but they are a dying breed. That's because the Internet isn't like a book at all; it's something quite new and different. The rise of 'search', and the company that came to dominate it, Google, seems obvious now because it has become so familiar. Search is a quick, easy, and efficient way of finding information. That's why Yahoo! is still merely a brand name and Google has ascended to the next level: it has become a verb.

What held back Yahoo! (and other companies) wasn't their lack of knowledge – they were, and still are, filled with talented people – it was what they already knew about the book. They were restricted by the same habitual thinking as the participants in the two-string problem. A similar principle operates in many other areas of creativity: film directors keep making (roughly) the same films, writers are writing the same books, photographers are taking the same pictures and composers composing the same music. They do this partly because more of the same sells, but also because more of the same is easier to generate, since the habit is already learnt.

Although the benefits of expertise on creativity appear obvious – more experience, knowledge, and technical skill – the case is far from open and shut. Early Gestalt psychologists, known for their slogan 'The whole is greater than the sum of its parts', thought knowledge could hinder the ability to solve insight problems. They pointed to experts' inability to escape the confines of their own habitual ways of thinking. Experts are good when new problems follow similar patterns to old ones, but can become blocked when they don't. When you have a hammer in your hand, everything looks like a nail. This sort of fixation often arises because the expert uses a familiar solution to start off with, and then becomes stuck in the same rut with all subsequent attempts to solve the problem. A large number of studies show that expertise often doesn't seep into related areas, and expertise can even inhibit performance when the task requires experts to ignore things they've learnt previously.[3]

A neat demonstration of this, by Jennifer Wiley at the University of Pittsburgh, had people trying to solve word puzzles.[4] Participants were given words like 'fly', 'boy', and 'bearing', and then asked what other word could be put either before or after to create three new words or phrases. The answer for this one is 'ball'. These baseball-themed answers continued throughout the word puzzles. The twist was that participants were given clues suggesting the answer had something to do with baseball, when sometimes it didn't. For those people whose knowledge of baseball was high, the misleading clues had more of a detrimental effect than for those whose knowledge was low. In the end, those with higher levels of knowledge about baseball actually performed worse on this task, which is exactly the opposite of what you'd expect. What was happening was that their knowledge was getting in the way. Participants with baseball expertise were biased in their first answer, which sometimes had nothing to do with baseball.

Much the same can happen when experts try to be creative in many different fields. As creative people acquire more experience and become more technically accomplished, their habitual ways of thinking and working tend to build self-imposed limits. A great example is provided by Japanese researchers who studied a master Chinese ink painter who had been working in the medium for many years.[5] Non-artists were asked to make some random marks on the paper before he started. The resulting paintings were then compared with similar artworks he made without the random marks. Despite his great skill and technical ability, he was able to create a new, livelier style around the random marks. The random lines seemed to get his mind moving in a novel direction. Using randomness might not work in many creative endeavours, but the general lesson the story tells about creativity is universal. Practice makes perfect, but it also makes the same thing over and over again.

None of which is to say we shouldn't practise our expertise – of course we should. Most highly creative people are experts in their field. The ink painter wouldn't have been able to create such beautiful pictures without his technical ability; neither would Einstein have been able to reshape physics without understanding the mathematics. It is virtually impossible to make creative leaps without existing knowledge to build on. It allows new problems to be compared with old, for the problem to be structured more effectively, and for likely methods of creative resolution to be identified. On top of that, experts typically think in a more abstract way about problems, and, as we'll see, this is a handy trick.

Advice for novices in every area of creativity is the same:

steep yourself in knowledge and keep working on the required skills and habits. The first lesson for anyone who wants to write is: do a lot of reading. It goes without saying that it's hard to be a creative cosmologist if you don't know anything about cosmology; or to paint a decent likeness of a tree if you've never picked up a brush in your life. It will depend on your particular area of creativity, but there are often some habits or sets of knowledge that you have to acquire before you can get creative.

Just be aware that, as studies have shown, expertise has its own blind spots, cul-de-sacs, and unexplored areas into which experience doesn't travel. Everyone already knows the importance of expertise; what few realize is how expertise and its attendant habits cramp creativity. The biggest danger for the expert is functional fixedness, getting stuck in a rut without realizing it.

This leaves us with a question: how do both experts and non-experts alike escape habitual ways of thinking to explore new pastures?

~~~~~~~~

If you want to see lightning-fast creativity in action, then the improvisational TV programme *Whose Line Is It Anyway?* isn't a bad place to start. The format is as simple as it is famous: professional improvisers make up funny scenes on the spot based on suggestions from the studio audience. Performers incorporate different events, styles, and emotions into their improvisations, and the results are frequently hilarious. The humour relies heavily on the skill of the performers in
~~~~~~~~

creating flights of fancy, but it wouldn't work without the apparently minor contribution of the audience. Without their random suggestions, the programme would fall apart, and this helps reveal an important lesson at the heart of creativity.

Most of us have experienced that feeling at the start of a project when, no matter how hard we try, nothing comes to mind. Similarly, writers often talk about the horror of the blank page or the cursor blinking on an empty screen. That's because when anything is possible, when anything could be done, sometimes nothing is possible and nothing gets done. Contrast that with the days when you have a list of specific, well-defined tasks which, once completed, allow you to relax in the evening reflecting on several jobs well done. The problem is that creative tasks are rarely well defined or specific; instead, they can prove about as easy to grasp as a rainbow.

Because we tend to think that creativity is thinking without constraints, we let the mind float free, and it promptly floats off for a nap. Rather counter-intuitively, psychologists have found that under the right circumstances, creativity can be increased by introducing constraints. In one early study, participants were split into three groups and asked to come up with a new invention from a set of raw materials.[6] One group was given a list of categories, such as 'vehicle', 'toy', and 'appliance', while another group was given specific parts, such as a ring, a wheel, and a tube. The third group, though, was allowed to use both the specific parts and the general list of categories. It was this third group, the one that had the most options, that produced the least creative inventions. Constraints, then, seem to help the creative process, and stricter constraints can make us even more creative.

We tend not to notice how many creative tasks benefit from constraints because they are built-in and have become invisible. For example, almost all popular music is in 4/4 time, four beats in the bar, with the emphasis usually landing on the first beat. Tracks are normally three or four minutes in length, contain a chorus, and so on. These are just a few constraints of many, and yet look at the variation that can be achieved. Many songs break these rules, but they often achieve their effects because there is a rule to break in the first place. Painters, writers, artists, and so on are all influenced by previous styles to various degrees, and it's these previous styles that provide constraints. The very limitations we impose on ourselves can be the seeds of our finest creations.

Psychologists have found that some of the most productive constraints, those that enable us to throw off habitual patterns of thought, are those that imagine a different world – that ask, 'What if?' For example, what if we abolished money? Or, what if we turned the Internet off? Or, what if psychologists ran the world? Although these might seem like ridiculous questions, when the thought is entertained, some fascinating new ideas spring to mind.

'What if' type questions, or 'counter-factual statements', can be used to help us escape from habitual ways of thinking. Research led by Keith Markman from Ohio University has studied counter-factual thoughts in a series of experiments.[7] They found a vital distinction between additive and subtractive mindsets. Additive mindsets focus on the addition of something to a situation; for example, say you are wet after getting caught in a shower; you might wish you'd had an umbrella. Subtractive mind-sets take something away: in this case, you might imagine a world where it hadn't rained.

It might seem like a very subtle distinction, but across three experiments, the research found that each promoted different ways of thinking. The additive mindset encourages people to think in a more expansive way, allowing them to generate more ideas. On the other hand, the subtractive mindset makes people think in a narrower, more analytical way, focusing their minds down on to the relationships between the problem's components. Both styles are vital in creativity, but at different stages. Sometimes, it's necessary to think expansively while looking for new components and connections, while at other times, typically later on, it's about working out how to fit the components together to make a working solution.

Returning to *Whose Line Is It Anyway?*, it's easy to see how the format uses constraints and 'what if' questions to break the performers out of old habits. What the contestants don't have, though, is time to come up with better ideas. We laugh at improvisational comedy partly because we know it's just been made up. Our standards are much higher for scripted comedy. Creative solutions to difficult problems need time to gestate, and psychological research agrees. Unfortunately, our first instinct is to follow a path we've explored before: habitual thinking once again. This can be a mistake.

The classic study on creative preparation, conducted by Jacob W. Getzels and Mihaly Csikszentmihalyi, asked art students to create a still-life painting of an object which was later professionally evaluated.[8] The study found that the students judged to have created the best work were those who spent the longest preparing – thinking about the object itself and how they were going to use it. When Mihaly Csikszentmihalyi returned to the same people at 7 and 18 years later, he found that it was these measures of problem identification and construction that predicted the artist's long-term success. Even 18 years later, artists who spent longer constructing the problem were more successful. This research, along with other findings, not only suggests that constraints can benefit the creative process, but also that we need to give ourselves time to analyse the problem.[9]

Unfortunately, the temptation with creative problems is to use habitual responses to get started on the solution immediately. Since problem-construction feels like a waste of time, though, it may be the most important part of the creative process. The choices made in the early stages have a massive impact later. That's why spending longer thinking about the problem before you dive in is likely to lead to higher levels of creativity in the final product. Like the art students painting a still life, when the options are almost infinite we need time to ponder the possibilities, and we are likely to do better if we take time to consider them. Fools rush in where the more creative dare to tread.

~~~

There is something magical about the power of opposites. The psychologist Albert Rothenberg first demonstrated the connection between creativity and the ability to generate opposites when, in the 1980s, he gave a word-association test to 22 winners of Nobel prizes in science. The Nobel laureates produced many more opposites, such as dark and light or bold and afraid, than a comparison group of students. Similarly, amongst the students, those rated as highly creative also produced more opposites than those with low creativity. Rothenberg thought this showed an essential truth about creativity, one that he continued to explore.

Over about 2,500 hours, Rothenberg interviewed 375 creative people from across the sciences, literature, and the arts, hoping to find the key to the creative process.[10] When his conclusions were published in 1996, Rothenberg named his main finding after the Roman god Janus. Busts and depictions of Janus are easy to spot because he's often shown with two faces, one bolted onto the back of the other, so that they point in opposite directions. If there is such a thing as a good creative habit, then Janus would be its poster boy.

Janusian thinking is all about the ability to conceive of opposite ideas, like those Janusian busts looking in both directions. It was Janusian thinking that pointed Albert Einstein towards his famous discovery of the general theory of relativity. In a thought experiment, Einstein imagined a person falling off a house and dropping a pebble out of his pocket. He realized that, while falling, the pebble would
~~~

remain stationary compared to the person. That person, while staring at the pebble and blocking out the rapidly approaching pavement and the wind screaming in his ears (this is a thought experiment, after all), could theoretically consider himself stationary. It's a ridiculous contradiction: how can someone be in motion, but also stationary? Nevertheless, it's this contradictory image that led Einstein to one of the most important breakthroughs in modern physics: how gravity works.

Rothenberg found this creative habit repeated, again and again, in different areas of creative endeavour. To achieve their breakthroughs, innovators typically went through four phases. They started with a very strong drive to create: without that, it's hard to push new ideas through. Next, departing from well-worn pathways of thought requires the conception of two opposite ideas, or opposite collections of ideas, like the two poles of a magnet. Third, the realization that the two poles can be integrated, and fourth, the full construction of the idea.

Assuming you're motivated, the first problem for any creative goal is coming up with the concepts to combine. Psychologists have found that using analogy is one handy way of finding concepts to set up in opposition; unfortunately, good analogies are hard to come by. Think about Einstein's vision of a man falling off a roof; it seems simple once you've heard it, but taken in the context of the highly complex problem he was facing, it was a masterstroke.

The key is envisaging the problem in a way that makes it easier to pick out the analogies. In one study, participants were asked to read about the behaviour of fictional people (or objects), and then try to draw analogies between them.[11] The information given, though, was represented in different ways: sometimes objects were described in general ways, at other times in specific ways. For example, a horse can be described as a domesticated, odd-toed ungulate mammal (specific) or just as a vegetarian (general). What researchers discovered was that analogical leaps were easier when problems were described in looser, more generic terms: then the participants' ability to pick out analogies increased by more than 100% on some of the tasks. So one way to encourage analogical thinking is to re-represent problems in more general, abstract terms. This research is just one example of a whole series of studies that show the importance of finding a problem's underlying structure. Focusing on the gist of the problem rather than its specific details makes it much easier for the mind to avoid habitual modes of thought and leap across the analogical gap to other types of problems in order to create novel solutions.[12]

One aid to seeing deeper structure is changing words which are specific to the problem into more general ones. Another technique used in psychological research is cutting out concrete concepts from the problem. Stripped of the surface details, the deep structure becomes more obvious and analogical leaps are more likely to take place. Mentally zooming out has the effect of reducing the space between stepping stones across a pond. And the closer together they are, the easier it is to get from one side to the other.

These techniques work in the real world as well. Businesses face the problem that habitual patterns of thought only create the same products that everyone else is marketing. The use of analogical reasoning in industry was examined by Oliver Gassman and Marco Zeschky in a study of breakthroughs in four engineering firms.[13] The firms manufactured all kinds of products, from skis, sewing machines, and valves through to components for car safety systems. Each was looking for ways of surmounting technical problems with their products. For example, the ski manufacturer was having problems with vibration at some speeds. Unfortunately, the company couldn't come up with a suitable solution. It wasn't until they started searching with three very general words in mind – 'damping', 'cushioning', and 'vibration' along with a frequency range – that they hit on a solution. The answer came from an inventor who had found a way of making bowed instruments sound better. Suddenly, the engineers saw the connection between violin bows and skis. The solution – applying an extra, thin layer – is now used in virtually all skis to damp down vibration.

Despite being in different industries, the three other firms each used similar analogical processes to attack their problems. The best solutions were achieved when the problem was represented at maximum abstraction, zoomed right out. This made it easier to see connections with other problems which had already been solved. At the same time, it made the habitual solutions less salient. Once again, abstraction brought the next stepping stone within reach, bringing what had been distant analogies closer together, so that the parallels were easier to see.

Now, take a moment to fix your mind on the idea of love. Think about what it means to you, what types of associations it brings to mind, how you feel about it. If you try to describe in words what love means, I'd be surprised if you didn't find it very difficult; it's certainly a lot harder than describing sex. That's because love is an abstraction like hope or ethics, something that's difficult to picture, whereas sex is concrete and can be pictured quite easily. Psychologists have found that this distinction between abstract and concrete ideas is crucial to escaping habitual ways of thinking.

The concepts of love and sex were used in a study that probed the roots of creativity.[14] Would it be possible, they wondered, for someone who was in an abstract frame of mind to be more creative than someone who was in a very concrete, nuts-and-bolts frame of mind? To test this out, they divided 60 participants into three groups and had one group think about love, another about sex, and a third about neither (the control group). Then they were all given some problems that required creative insight and some other problems that merely required tried-and-tested analytical methods.

The results showed that those who had been thinking about love performed better than the control group on the task that required a creative insight. On the other hand, those who had been thinking about sex did worse than the control group on the creative-insight problem. When it came to analytical thinking, though, the results were reversed. Those with sex on their mind were better at thinking analytically, while those with love on their mind were worse. This finding isn't just about love and sex, though, tempting as both are to think about. They are examples of a deeper structure: the balance between abstract and concrete thinking.

This balance can also be influenced by how we think about time. When we think about the future, it tends to encourage abstract thought, while things which are psychologically close in time are perceived in more concrete terms. For example, if you're taking a trip tomorrow, your mind will focus in on whether you've packed your bags, got your tickets, and worked out the route. On the other hand, if you are planning a trip in six months, the mind is encouraged to roam across more abstract concepts, like the history, culture, and cuisine of your destination. Distance in time tends to open up your thought processes to more wide-ranging possibilities.

In six experimental tests of this idea, people's perspectives on time were manipulated.[15] Participants were asked to carry out tasks requiring creative insight, but first to imagine that they were doing it in a year's time for a distant/abstract group, or tomorrow for a close/concrete group. As expected, it was those who sent their minds off to the distant, abstract future who shone in these tasks compared to those stuck in the concrete world of the near future. One of the experiments demonstrated that participants didn't even need to imagine they were doing the task in the future; all it took was for them to think about their own lives in one year's time. The disadvantage of abstract thought, however, like concentrating on the concept of 'love' in the previous study, was that it tended to reduce performance at analytical problem-solving.

If distance in time can cue an abstract state of mind, then perhaps distance in space can do the same. One study asked participants to attempt a creative-insight task, which some were told had been developed at the local US university, and others were told had been developed in Greece.[16] Incredibly, this was enough to cue up an abstract state of mind: participants had more creative insights when told the test came from Greece. This 'thinking at a distance' led to more fluency, flexibility, and originality.

If you're stuck in a rut with a creative problem, going down the same old avenues of thought, then these studies suggest a way out. Whether it's using spatial or temporal distance, the key is always inducing psychological distance in order to get into an abstract state of mind and away from established habits. People often say while they are fruitlessly searching for insight into a problem that they are 'too close to it'. In a psychological sense this turns out to be true. For creativity, the further away, the better the view. At least, it is at the start of the process, when new ideas and originality are most important. Later on, different types of thinking come into play.

~~~

In 1865, the German chemist August Kekulé made his greatest scientific breakthrough, publishing a paper on the structure of benzene, a problem that had frustrated early organic
~~~

chemists for decades. He later described how the insight had come when he had given up work and turned his chair to the fire, falling into a reverie.[17] As often happened to him in this state between waking and sleeping, he began to visualize the atoms dancing before his eyes. This time, though, the lines of atoms appeared like snakes, and suddenly one of these snakes curled up to bite its own tail. When he awoke, he realized the dream was telling him that benzene had a circular structure, with six carbon atoms forming a hexagon. This was the founding discovery in what is now an enormous field: the chemistry of ring compounds.

Although the idea that Kekulé drew inspiration from a dream has been questioned,[18] it is one story of creativity that praises the defocused mind and highlights what psychologists have discovered in many studies: that a wandering mind is often associated with increased creativity. This echoes the benefits of abstract thinking and the notion that great ideas spring from the combination of two previously disparate concepts. Since the mind wandering away from habit is more likely to wander into new ideas, it has a better chance of success when trying to combine them.

There's no better example of the wandering, playful, creative mind than that of a child. If you watch children play, they can pick up anything and imagine it as something else. Boxes become houses, tunnels or trees, while carrots are spaceships and your favourite shoes are plant pots. They haven't yet built up habits of thought that limit their imagination, so things aren't yet stuck in one category or another. As a result, children operate free of many of the constraints and worries which constantly trouble grown-ups and they have a natural facility for counterfactuals or 'what if' scenarios.[19] It's no surprise, then, that they can be incredibly creative, without even knowing the meaning of the word.

So what if, when people try to solve problems creatively, they are encouraged to think like a 7-year-old child? This is exactly the logic followed by researchers who primed half of their participants to think like 7-year-olds, before giving them a creativity test.[20] Priming is a technique favoured by psychologists when they want to unconsciously put people in a different frame of mind, like when we smell freshly baked bread and our minds turn automatically to lunch. In this case, though, participants were asked to write about what they would do if they had the day off, but only half of them were asked to do it as though they were 7 years old. The results showed that adults who had been primed to think like a child scored higher on a creativity test compared with those who had not. This study suggests that something as simple as imagining yourself as a child has the power to boost playfulness, openness to experience, and so, creativity.

But while playful, defocused attention stimulates creativity, controlling attention is also important. This is simply because, as Thomas Edison once memorably said: 'Genius is 10% inspiration and 90% perspiration'. Controlled attention is what children generally lack, and that's why they're not in charge of our creative industries. Without analysis, evaluation, and persistence, there's little hope that creative tasks will ever be finished. Backing up this idea, psychologists have found that being able to focus on a problem is associated with high levels of creativity. We need to be able to work out which solutions might have a chance of working and how ideas can be translated into reality. Kekulé's snake dream and Einstein's falling man are attractive images, but they are nothing without the subsequent hours of painstaking, focused work to back up the original flash of inspiration.

Psychologists, then, seemed to be faced with an apparent contradiction at the heart of creativity. On the one hand, those with wandering, defocused, childlike minds seem to be the most creative; on the other, it seems to be analysis and application that's important. The answer to this conundrum is that creative people need both the wandering *and* the controlled mind; it all depends on the stage. Studies of creativity in scientists show that they playfully explore a wide range of ideas at first, but once the problem is well defined, they focus in on the most useful concepts to the exclusion of others. The key to creativity is being able to switch between a wide-open, playful mind and a narrow analytical frame.

This ability to switch between different ways of analysing a problem has been examined in an experiment in which participants took tests of creativity and flexible thinking.[21] Those who scored higher on the creativity test also displayed more flexible thinking. While those low in creativity could still move from one type of thinking to another, they weren't as quick as those who measured high in creativity. While still a relatively new finding, results pointing in the same direction have been obtained by psychologists using different tasks in other laboratories. The same pattern repeats: more creative people find it easier to switch between strategies.[22]

A creative habit for attention, then, means finding a balance between the playful, wandering mind and the focused, highly analytical mind. Not all of us are blessed with the ability to think flexibly. Some people are great at buckling down and focusing on the nitty-gritty, while finding it difficult to be playful. Others are playful all the time, but find it hard to evaluate and implement their ideas. You probably know which category applies to you and which needs work.

Many great creative geniuses over history have identified their weakness and addressed it. Often, it's distraction. Charles Darwin was notorious for his requirement for solitude and had to work in complete seclusion. Novelist Marcel Proust had his bedroom lined with cork, and the philosopher Arthur Schopenhauer noted that he, like other great minds, required silence to avoid having his thoughts interrupted. So if your mind wanders when you should be analysing the details of your problem, then don't worry, you're in good company. Just remember that all these great minds had to find a way to balance their playful and analytical sides in order to develop truly creative habits.

13

HAPPY HABITS

In one episode of *The Simpsons*, Bart has a sudden premonition about what life has in store for him.[1] He wakes up, sighs, and says: 'Monday. Here we go again.' At breakfast, his father Homer hogs the orange juice and Bart looks bored. On the school bus, his sister Lisa plays her saxophone and Bart looks bored. At school, he's shot in the back of the head by a pea shooter and he looks bored. On the way home, he's chased by bullies while looking bored. Then, at home, he settles down to watch TV, still with the same bored expression on his face.

Bart is still contemplating a life of tedious repetition when he sees an ad on TV for an exciting cruise. After talking his parents into it, the whole family finds themselves having the time of their lives, enjoying new surroundings and activities. Soon, though, Bart realizes the cruise will be over and he'll be back to his old dreary, repetitive life again with its routine, emotionless activities. The episode wraps up with a consoling message from Lisa, Bart's sister. She tells him that it's true that life can be boring and repetitive, and that moments of true joy are frequently few and far between, but that we've got to appreciate the good moments as best we can when they do come.

Lisa's advice sounds clichéd because it's as old as the hills. The Greek philosopher Epicurus, who was much concerned with happiness, would have enjoyed this episode of *The Simpsons* because it chimes with one of his central teachings. As a so-called 'hedonist' philosopher, Epicurus is sometimes wrongly thought of as supporting overindulgence, especially of food. But far from endorsing a pleasure cruise, Epicurus would have known that Bart was setting himself up for disappointment. In fact, his philosophy was about living sparingly and appreciating small pleasures wherever we can. He once said: 'Give me plain water and a loaf of barley bread, and I will dispute the prize of happiness with Zeus himself.' Even while dying painfully of kidney stones, he was writing letters to friends telling them how much he appreciated their friendship and how cheerful he was about his philosophical contemplations. For Epicurus, the goal of living a happy life was inseparable from living a virtuous life.

One of the goals of habit change is to make ourselves happier. We might be trying to improve our work habits so that we can get more done in less time, or our socializing habits so that we can spend more time with friends and family. Good habits can do all sorts of things for us, but will they make us happy? Both Epicurus and Bart Simpson found that understanding what gives us pleasure and how to live a happy life are no simple matter. The 'how' of happiness is the problem because it can be difficult to say exactly what brings happiness. We have, however, found out much about how happy habits operate that would have pleased Epicurus and we know more about why, like Bart Simpson, we can find taking pleasure in our everyday activities so difficult.

~~~~~~~~~

It's one of the great paradoxes of life that we all want to be happy, yet sometimes it's hard to know exactly where happiness comes from. Part of the reason is that essentially we have no control over around half of our overall levels of happiness.[2] We certainly go up and down as the days and weeks pass, but ultimately we tend to return to the same levels. Like many other aspects of our bodies and minds, our overall happiness is partially set by our genes.[3] While these do interact to a certain extent with the environment, on a day-to-day basis this 50% is difficult to budge.

Of the remaining 50%, about 10% of how happy (or not) we feel seems to be down to our circumstances: this includes things like income, education, age, being married, residence, and background. Considering how much importance is usually attached to circumstances, this might sound like a
~~~~~~~~~

frighteningly small amount. People work long hours to improve their lot in life and to give their children a better education, and so on. However, while improved circumstances probably increase our satisfaction with life (how we rationally evaluate it), the influence on how we feel is relatively small. This helps explain why a doubling in US income in the last half-century has not led to any increase in overall levels of happiness.[4]

One important reason for this is habituation. Human beings are wonderfully adaptable and we adapt to our new circumstances, whether better or worse, with frightening speed. Studies have shown that when we rationalize events, it helps reduce their emotional impact.[5] This process of making meaning out of the things that happen to us is automatic and at least partly unconscious. But along with these explanations is also the idea that as our new circumstances produce habits, their performance becomes unconscious. We stop noticing and taking pleasure from positive changes in our circumstances because they've been incorporated into our automatic routines.

An early study which hinted at this compared people who'd won up to $1,000,000 in the lottery with their less fortunate neighbours.[6] Six months after their windfall, winners rated their happiness much the same as before they'd won. They seemed to have adapted to their circumstances, and in many ways the excitement of winning the lottery had taken the gloss off other, more mundane activities. Just as we quickly get used to positive changes in circumstances, we also show surprising psychological resilience to negative changes in circumstances. This is both a blessing and a curse. While we may be glad to recover from life's tragedies, it's irritating that when we do succeed, the fruits of our labours are soon taken for granted. The same effect is seen to be true about money. When people get pay rises, they are happier to start off with, but soon adapt to their improved circumstances, leaving them not that much happier in the long term.[7] It's also true of countries as a whole, which don't get much happier with increases in GDP.[8]

The idea of habituation can be intuitively hard to believe. When we project ourselves forward into the future – say, dreaming that we win the lottery next week – it seems difficult to imagine that we will get used to these exciting new circumstances. Part of the problem seems to be a weird glitch in how we estimate our future emotions. We're very good at predicting our emotional reactions to events in general: we know that floating feels better than drowning and flying feels better than crashing. What is harder is predicting just how good (or bad) things will make us feel and for how long the feeling will last. On a recent flight, I sat near a group of teenagers, most of whom were obviously flying for the first time. They found everything exciting: take-off, looking down at the Alps, turbulence, landing – even the safety instructions. Could those adolescents ever imagine that in the future all these things will become boring? This is what psychologists have called a failure in our 'affective forecasting'. Although we habituate quickly to the ups and downs of life, and it happens time and time again, we don't seem to learn from our experience.

Take one study led by Daniel Gilbert of Harvard University that asked people to predict how all kinds of negative events would affect them.[9] Participants were asked to imagine how they would feel about things like the end of a romantic relationship, reading a story of a child's death, receiving a job rejection and so on. Time and again, the same pattern emerged: people predicted they would feel worse than they actually did. People make exactly the same errors when they try to predict how they will react to positive events as well. Although a promotion, a new car, or a new relationship might make us happier for a period, we soon get used to it.

If we keep overestimating the effect of these events and then keep adapting to the changes, why don't we learn? The simple explanation is probably the best: we forget. We don't recall our past predictions of how we'd feel, instead replacing them with how we actually feel now.[10] Our emotions are always anchored in the present, comparing everything that might happen with how things are now.

When you think about it, the idea that changes in circumstance do relatively little to affect our happiness is good news. This is because it's hard to make very significant changes to our jobs, families, income, and all the rest. There's certainly nothing we can do about our genetic set-point. After we remove our genes and our circumstances, though, we're left with what we do every day, which includes our habits. For example, you have a walk in the park, you feel better; you sit in traffic during your commute, you feel worse; you put on your favourite music, you feel better... and so on. Put crudely, if you add up all these little ups and downs, what emerges is your level of happiness within the limits set by your genes and your circumstances.

All this means that we have a certain level of control over our happiness if we choose the right habits, but which habits should we cultivate? You will have certain habits which routinely make you feel happy – perhaps it's origami, gardening, listening to blues music, or doing nothing at all. These are fine, but psychologists have looked for more broadly acceptable habits: things that will work for most people.

Many different activities have been tested by psychologists and some have been shown consistently to improve how we feel day to day. These include things like boosting positive thinking and social connections, dealing with stress, being present in the moment, committing to goals, and practising gratitude. The psychologist Sonja Lyubomirsky describes them in detail in her excellent book *The How of Happiness*.[11] Let's take practising gratitude as an example, though, as it illustrates a problem with happy habits, and indeed, with the pursuit of pleasure in general.

The idea of practising gratitude is now a relatively well-known exercise and is also much researched. As with many of the studies described in this book, the effects of a particular thought or behaviour are compared with a control group. For example, a typical study of gratefulness has participants assigned to one of three groups:

1. The first group are asked to write down five

things they are grateful for that have happened in the last week.

2. The second group are asked to write down five daily hassles from the previous week.
3. The third group simply list five events that have occurred in the last week, but are not told to focus on positive or negative aspects.

All three groups carry out their assigned activity for ten weeks. In a study using this design, the types of things people listed in the 'gratitude' condition included a sunset through the clouds, the chance to be alive, and the generosity of friends.[12] In the hassles condition, people listed things like taxes, difficulties finding parking, and burning their macaroni and cheese.

The results of the study suggested that the gratitude condition was surprisingly powerful. People who'd focused on their weekly uplifts felt happier and appeared happier to their friends than those in the other two groups. They were also more optimistic about the future, felt better about their lives, and even did almost 1.5 hours more exercise a week than those in the hassles or events conditions.

There are all sorts of reasons why practising gratitude works. Just a few of these are that it encourages us to think about our social connectedness, helps us savour and enjoy everyday life, and makes us more habitual positive thinkers. Lyubomirsky advises that people should use whatever method for practising gratitude works best for them: some people find that keeping a gratitude journal is useful; others prefer just to think about the things for which they are grateful.

Unfortunately, there's rather a large and ugly fly in the ointment. That fly is habituation. Any activity we carry out to try to increase our happiness, if it eventually becomes completely routine, can soon be unconscious and unnoticed. Imagine hearing yourself droning in a bored monotone the mantra: 'I'm grateful to my partner for their love, to my employer for the job… ' All the while, part of you is wondering what's for lunch and how much longer all this gratitude malarkey is going to take. With repetition over the years, this is not going to increase happiness much. And yet it is a natural consequence of developing a habit. It's the same reason that seeing your one-hundredth superhero movie is unlikely to be as captivating as that first one. For some habits, the fact that they become automated and unconscious doesn't matter; indeed, it's a bonus. While I'm flossing, I'm happy to spend the mental time elsewhere. But there are plenty of other routine activities that, while we want to keep doing them regularly, we don't want the spark to go out.

That's not all the bad news: here's why the fly is so big and hairy. I mentioned before that people adapt to changed life circumstances, whether good or bad. There's another little wrinkle: it seems that we adapt to positive experiences more quickly than negative ones.[13] In other words, we lose the pleasure from good habits more quickly than the pain from bad habits. This is extremely irritating, but not impossible to overcome. According to the research in positive psychology, one way in which we can fight back against automatic adaptation to pleasurable experiences is by adhering to that old saying that variety is the spice of life. If we can vary our happy habits enough, rather than repeating them in the same way over and over again, we can continue to reap the rewards. This does require a little conscious thought and creativity. For example, I'm a big fan of cycling, and although I cycle regularly, I try to vary the routine as much as possible. One day, I go clockwise around the park; the next, anticlockwise; the next, I travel in a figure of eight; the next, I avoid the park completely. When the weather (and sunlight) allows, I ride at different times of day. I also try to vary the music I listen to and stop for breaks in different places. In other words, I do everything to create little variations in the regular routine.

Introducing these sorts of variations into your habits, including those designed to increase happiness, can be effective in reducing the effects of habituation. This has been tested in a study that had participants make small changes to their lives, like joining a club or sports team, and report on how much variety this introduced.[14] When their happiness levels were measured afterwards, those who performed activities with high levels of variety and high awareness showed the largest increases in happiness.

The same is true of all the happiness-enhancing activities that research has found to be effective. Another example of a happy habit is visualizing your best possible self. Carrying out this exercise typically involves imagining your life in the future, but a future where everything that could go well, has gone well. This may sound like an exercise in pure fantasy, but crucially, it does have to be realistic. You imagine having reached those realistic goals that you have set for yourself. Then, to help cement your visualization, you commit your best possible future self to paper. This exercise draws on the proven benefits of writing expressively about your innermost thoughts and feelings.[15]

A study which tested this activity involved participants writing about their best possible future selves for twenty minutes over four consecutive days.[16] This group was compared with three others: one writing on a neutral topic, one writing about traumatic life events, and another writing about both traumatic events and their best possible future selves. The results showed that those who had only written about their best possible selves showed greater improvements in subjective well-being compared to all the other groups. The benefits of the exercise could even be measured fully five months later. Other studies have also shown benefits over longer periods.[17]

Once again, though, to maintain the gains in well-being from this happy habit, variation is required. Write on different days, at different times, about different subjects, sometimes for a short period, sometimes for longer. Sit in a different chair or in the park or on the train. Or don't sit at all. Only, make sure you fight the habituation.

Another happy habit which has been shown to increase well-being is savouring. A bad habit we get into is thinking that all the good things in life are to come in the future. Clever, industrious, conscientious, successful people often tell themselves they are working hard for what is to come. In some respects, this is an excellent habit because it encourages us to commit to long-term plans and to avoid dangerous temptations in the moment. But your focus can easily switch too strongly to the future. In other words, you can end up sacrificing your pleasure in the moment for some imagined future that never arrives. Or worse: it has arrived and you haven't noticed.

The happy habit of savouring is simply reining your mind back in and forcing it to focus on the good things in life, to 'stop and smell the roses'. People do this naturally, but four methods have empirical support: showing your emotions, being present in the moment, celebrating positive events with others, and positive mental time travel. They are really ways of focusing the mind on thoughts that are likely to produce positive emotions. Unfortunately, people also naturally practise four opposing, dark, unhappy habits. Instead of showing our emotions, sometimes people don't like to show they're happy. Whether it's because of fear, shyness, or modesty, people do hide their positive emotions. Instead of being present, or enjoying what's happening now, our minds have the habit of wandering off. Unfortunately, we quite often wander off to our worries. This dampens down the positive emotion we feel. Instead of celebrating success, sometimes people have a habit of looking for faults. Yes, they say to themselves, this was good, but it could have been better. This tends to reduce life satisfaction, optimism, self-esteem, and happiness. Finally, the other side of mental time travel is that our minds can just as easily take us back to past embarrassments or forward to future irritations.

All eight strategies, four positive and four negative, have been compared by asking people to imagine how they would react to pleasurable experiences such as having a romantic weekend away and finding an incredible waterfall while out hiking.[18] The results showed that positive mental time travel and being present were most strongly associated with heightened pleasure. As for satisfaction with life – our evaluation of how we're doing – the best savouring strategy was celebrating wins; according to this research, there's nothing better than this for helping us feel our lives are going well. On the other hand, fault-finding and letting the mind wander to negative events are most likely to reduce our satisfaction with life.

This research can't tell us when to use these happy habits, although it seems that variety is important. Indeed, two of the 'good' strategies are opposites: one involves focusing on the here and now, while another involves drifting off somewhere else. What they did find was that the people who were most flexible about which strategy to use experienced the most pleasure. It's likely that if you typically use one of these strategies already, habitually practising a different one will increase your flexibility, and in turn both your well-being and life satisfaction.

These are just a few examples of the types of habits that can help cultivate greater well-being. Other happy habits with evidence to support them include 'committing' acts of kindness[19] and working out how to use personal strengths.[20] In common with any new habit, a happy habit also needs to fit with your personality and circumstances in order to be effective. If it feels wrong or you don't find it useful, then change it. The research is just suggesting which activities are best, on average, across many people – for you, it's likely that some things will work much better than others.

~~~~~~~~

The study of happy habits has something very important to teach us. It is a reminder that good habits we already have – and new habits we choose to break in – can easily become forgotten parts of our routines. Things we used to take pleasure from, like making a cup of tea or walking round the block, can, with time, become bland and emotionless. As automaticity increases, our experience of being in the moment recedes; we feel less alive, fail to notice the world around us, and become disconnected from our experience. Our minds travel off elsewhere, away from what we're doing and towards ruminating about the past and worrying about the future.

This was demonstrated in a study by Matthew Killingsworth and Daniel Gilbert.[21] They asked thousands of participants to track their happiness on a mobile-phone app. The app randomly interrupted them every now and then throughout the day and asked them three simple questions: (1) How are you feeling right now?; (2) What are you doing right now?; and (3) Are you thinking about something other than what you're currently doing? Almost half the time people were asked, at that moment their minds were wandering from whatever they were doing – 43% to pleasant
~~~~~~~~

topics, 27% to unpleasant topics, and the rest to neutral topics. The only time their minds weren't wandering was when they were having sex. The interesting thing was that both neutral and unpleasant topics, which comprised 57% of mind-wandering, made people considerably less happy than their current activity, whatever it was. Even when thinking happy thoughts, they were no happier than when fully engaged with their current activity. The most pleasurable was having sex, after which came exercising, then socializing, playing, listening to music, walking, and eating. At the other end of the scale, the least-happy habits were working, using the computer, commuting, and grooming.

This clearly shows us one of the dangers of habits. People's emotions are frequently detached from their habitual activities. The mobile-phone study is demonstrating the same thing. During the performance of habits such as commuting, doing housework, shopping, and eating, the mind tends to wander away. Although it could potentially wander away to a happier place, this doesn't seem to be what happens in practice. Instead, on average, people's minds wander away to thoughts that make them less happy. This means that the pleasure we get from an activity generally comes from being engaged with it, whatever it is.

Making or breaking a habit is really just the start. To develop a truly fulfilling and satisfying happy habit, it's about more than just repetition and maintenance; it's about finding ways to adjust and tweak habits continually to keep them new; to avoid mind-wandering and the less pleasurable emotional states that accompany it. There is great enjoyment to be had in these small changes to routines. When life is the same every day, it gets boring. We have to acknowledge and try to understand our habits, but also to rise above them, to continue working on how they can be changed, improved, or just tweaked.

Here, we're peering into an area that is only partly habitual. We might regularly perform activities in certain circumstances as a result of particular environmental cues, but then we should carry out the activity slightly differently each time. Imagine an opera singer about to sing the same aria she's practised for years and years. The cues to her singing are all the same: she is in the opera house, the audience is sitting quietly, listening, the orchestra is reaching the same point in the score, she opens her mouth to sing.... and yet what emerges is subtly different from before. She finds a new tone or rhythm, a slight nuance of meaning or shade that wasn't in previous performances. Something happens to pull the audience out of their own habits of thought, and into the moment, and suddenly they hear this piece they've heard many times before anew. It's the reason audiences attend live performances, and it's the reason singers go on singing: they're looking for something new in the familiar.

These ideas are stretching the formal definition of a habit which involves the same behaviour or thought in the same situation. For happy habits, we need slightly different behaviours in slightly different circumstances. We need the habit to rise above itself. Breaking and making new habits is just the first step. The ideal situation is an automatic initiation of the behaviour, but then a mindful, continuously varying way of carrying it out. A new type of hybrid habit: a mindful habit.

Of course, we have already developed many of these types of mindful habits naturally. When we get home, we might habitually look in the fridge for something to eat, but who knows what new concoction we might cook up? When we see a friend in the street, we habitually say hello and ask after their health, but who knows where the conversation will lead? Automatic, unconscious habits only set us off in the right direction; it is up to our conscious, willed, creative selves to decide where we are going and how we will get there. Habits can be a force for personal conservatism: if they take complete control, they can lock us into the same boring grooves. Habits can also set us free from the routine aspects of everyday life and allow us to realize our full potential. The challenge is to work out which habits keep leading to deadends and which habits lead to interesting new experiences, happiness, and a sense of personal satisfaction.

Where will you start?

... NOW READ ON AT PSYBLOG

Some of the ideas discussed in this book had their first airing on my website PsyBlog (www.psyblog.co.uk). I continue to write there regularly, as I have done for many years, about scientific research into how the mind works. Please join our conversation.

ACKNOWLEDGEMENTS

I owe the biggest debt to all of the scientists whose work this book relies on. It is their advances that have inspired me to think and write about habits.

For helping to make this book a reality in both the US and the UK, thanks to Luba Ostashevsky, Danielle Svetcov, Katie McHugh, John Radziewicz, and Robin Dennis.

Thanks to my sisters, Sarah and Alice, for that chat we had down the pub.

Love and gratitude, as always, to Mina for driving me forward.

NOTES

1. Birth of a habit

1 Maltz, M. *Psycho-Cybernetics A New Technique for Using Your Subconscious Power*. New York: Pocket Books, 1960.
2 Lally, P., C.H.M. van Jaarsveld, H.W.W. Potts, and J. Wardle. 'How Are Habits Formed: Modelling Habit Formation in the Real World'. *European Journal of Social Psychology* 40, no. 6 (2010): 998–1009.
3 Wood, W., J.M Quinn, and D.A. Kashy. 'Habits in Everyday Life: Thought, Emotion, and Action.' *Journal of Personality and Social Psychology* 83, no. 6 (2002): 1281.
4 Bargh, J.A., and T.L. Chartrand. 'The Unbearable Automaticity of Being.' *American Psychologist* 54, no. 7 (1999): 462.
5 Frijda, N.H. 'The Laws of Emotion.' *American Psychologist* 43, no. 5 (1988): 349.
6 Wood, W., L. Tam, and M.G. Witt. 'Changing Circumstances, Disrupting Habits.' *Journal of Personality and Social Psychology* 88, no. 6 (2005): 918.
7 Quinn, J.M., and W. Wood. 'Habits Across the Lifespan'. Unpublished Manuscript, Duke University, Durham, NC (2005).

2. Habit versus intention: an unfair fight

1 Lewicki, P., T. Hill, and E. Bizot. 'Acquisition of Procedural Knowledge About a Pattern of Stimuli That Cannot Be Articulated'. *Cognitive Psychology* 20, no. 1 (1988): 24–37.

2 Neal, D.T., W. Wood, P. Lally, and M. Wu. 'Do Habits Depend on Goals? Perceived Versus Actual Role of Goals in Habit Performance'. Manuscript Under Review, University of Southern California (2009).

3 Fishbein, M., and I. Ajzen. *Belief, Attitude, Intention and Behavior: An Introduction to Theory and Research*. Reading, MA: Addison-Wesley, 1975.

4 Triandis, H.C. *Interpersonal Behavior*. Monterey, CA: Brooks/Cole Pub. Co., 1977.

5 Ajzen, I. 'The Theory of Planned Behavior'. *Organizational Behavior and Human Decision Processes* 50, no. 2 (1991): 179–211.

6 Festinger, L. *A Theory of Cognitive Dissonance*. Stanford, CA: Stanford University Press, 1957.

7 Goethals, G.R., and R.F. Reckman. 'The Perception of Consistency in Attitudes* 1'. *Journal of Experimental Social Psychology* 9, no. 6 (1973): 491–501.

8 Ji, M.F., and W. Wood. 'Purchase and Consumption Habits: Not Necessarily What You Intend'. *Journal of Consumer Psychology* 17, no. 4 (2007): 261.

9 Ouellette, J.A., and W. Wood. 'Habit and Intention in Everyday Life: The Multiple Processes by Which Past Behavior Predicts Future Behavior'. *Psychological Bulletin* 124 (1998): 54–74.

10 Webb, T.L., and P. Sheeran. 'Does Changing Behavioral Intentions Engender Behavior Change? A Meta-analysis of the Experimental Evidence'. *Psychological Bulletin* 132, no. 2 (2006): 249–268.

3. Your secret autopilot

1 Johansson, P., L. Hall, S. Sikström, and A. Olsson. 'Failure to Detect Mismatches Between Intention and Outcome in a Simple Decision Task'. *Science* 310, no. 5745 (2005): 116–119.

2 Nisbett, R.E., and T. D. Wilson. 'Telling More Than We Can Know: Verbal Reports on Mental Processes.' *Psychological Review* 84, no. 3 (1977): 231.

3 Asendorpf, J.B., R. Banse, and D. Mücke. 'Double Dissociation Between Implicit and Explicit Personality Self-concept: The Case of Shy Behaviour.' *Journal of Personality and Social Psychology* 83, no. 2 (2002): 380.

4 Wilson, T.D., and E.W. Dunn. 'Self-knowledge: Its Limits, Value, and Potential for Improvement'. *Annual Review of Psychology* 55 (2004): 493–518.

5 Spalding, L.R., and C.D. Hardin. 'Unconscious Unease and Self-handicapping: Behavioral Consequences of Individual Differences in Implicit and Explicit Self-esteem'. *Psychological Science* 10, no. 6 (1999): 535–539.

6 Bosson, J.K., W.B. Swann Jr, and J.W. Pennebaker. 'Stalking the Perfect Measure of Implicit Self-esteem: The Blind Men and the Elephant Revisited?' *Journal of Personality and Social Psychology* 79, no. 4 (2000): 631.

7 Hofmann, W., B. Gawronski, T. Gschwendner, H. Le, and M. Schmitt. 'A Meta-analysis on the Correlation Between the Implicit Association Test and Explicit Self-report Measures.' *Personality and Social Psychology Bulletin* 31, no. 10 (2005): 1369–1385.

8 Lhermitte, F. '"Utilization Behavior" and Its Relation to Lesions of the Frontal Lobes.' *Brain: a Journal of Neurology* 106 (1983): 237.

9 Wilson, T.D., and D.S. Dunn. 'Effects of Introspection on Attitude-behavior Consistency: Analyzing Reasons Versus Focusing on Feelings'. *Journal of Experimental Social Psychology* 22, no. 3 (1986): 249–263.

10 Wilson, T.D., D.J. Lisle, J.W. Schooler, S.D. Hodges, K.J. Klaaren, and S.J. LaFleur. 'Introspecting About Reasons Can Reduce Post-choice Satisfaction'. *Personality and Social Psychology Bulletin* 19 (1993): 331–331.

4. Don't think, just do

1 Bargh, J.A., M. Chen, and L. Burrows. 'Automaticity of Social Behavior: Direct Effects of Trait Construct and Stereotype Activation on Action.' *Journal of Personality and Social Psychology* 71, no. 2 (1996): 230.

2 Shih, M., N. Ambady, J.A. Richeson, K. Fujita, and H.M. Gray. 'Stereotype Performance Boosts: The Impact of Self-relevance and the Manner of Stereotype Activation.' *Journal of Personality and Social Psychology* 83, no. 3 (2002): 638.

3 Dijksterhuis, A., and A. van Knippenberg. 'The Relation Between Perception and Behavior, or How to Win a Game of Trivial Pursuit.' *Journal of Personality and Social Psychology* 74, no. 4 (1998): 865.

4 Wood, W., and D.T. Neal. 'A New Look at Habits and the Habit-goal Interface.' *Psychological Review* 114, no. 4 (2007): 843.

5 Sheeran, P., H. Aarts, R. Custers, A. Rivis, T.L. Webb, and R. Cooke. 'The Goal Dependent Automaticity of Drinking Habits'. *British Journal of Social Psychology* 44, no. 1 (2005): 47–63.

6 Aarts, H., and A. Dijksterhuis. 'Habits as Knowledge Structures: Automaticity in Goal-directed Behavior.' *Journal of Personality and Social Psychology* 78, no. 1 (2000): 53.

7 Bargh, J.A. 'The Four Horsemen of Automaticity: Awareness,

Intention, Efficiency, and Control in Social Cognition'. *Handbook of Social Cognition: Basic Processes* 1 (1994): 1–40.

5. The daily grind

1 Rinehart, L. *The Dice Man*. New York: William Morrow, 1971, p. 12.
2 Avni-Babad, D. 'Routine and Feelings of Safety, Confidence, and Well-being'. *British Journal of Psychology* 102, no. 2 (2011): 223–244.
3 Wood, W., J.M. Quinn, and D.A. Kashy. 'Habits in Everyday Life: Thought, Emotion, and Action.' *Journal of Personality and Social Psychology* 83, no. 6 (2002): 1281.
4 Yin, H.H., and B.J. Knowlton. 'The Role of the Basal Ganglia in Habit Formation'. *Nature Reviews Neuroscience* 7, no. 6 (2006): 464–476.
5 Knowlton, B.J., J.A. Mangels, and L.R. Squire. 'A Neostriatal Habit Learning System in Humans'. *Science* 273, no. 5280 (1996): 1399.
6 Fiese, B.H., T.J. Tomcho, M. Douglas, K. Josephs, S. Poltrock, and T. Baker. 'A Review of 50 Years of Research on Naturally Occurring Family Routines and Rituals: Cause for Celebration?' *Journal of Family Psychology* 16, no. 4 (2002): 381.
7 Stinson, D.A, J.J., Cameron, J.V. Wood, D. Gaucher, and J.G., Holmes. 'Deconstructing the 'reign of Error': Interpersonal Warmth Explains the Self-fulfilling Prophecy of Anticipated Acceptance'. *Personality and Social Psychology Bulletin* 35, no. 9 (2009): 1165.
8 Buss, D.M. 'Toward a Psychology of Person–environment (PE) Correlation: The Role of Spouse Selection.' *Journal of Personality and Social Psychology* 47, no. 2 (1984): 361.
9 Werner, C., and P. Parmelee. 'Similarity of Activity Preferences Among Friends: Those Who Play Together Stay Together'. *Social Psychology Quarterly* (1979): 62–66.
10 This weak connection between actual similarity and attraction was also seen in a meta-analysis of 313 studies: Montoya, R.M., R.S. Horton, and J. Kirchner. 'Is Actual Similarity Necessary for Attraction? A Meta-analysis of Actual and Perceived Similarity'. *Journal of Social and Personal Relationships* 25, no. 6 (2008): 889.
11 Dainton, M. 'Maintenance Behaviors, Expectations for Maintenance, and Satisfaction: Linking Comparison Levels to Relational Maintenance Strategies'. *Journal of Social and Personal Relationships* 17, no. 6 (12 January 2000): 827–842.
12 Nelson, R.R., and S.G. Winter. *An Evolutionary Theory of Economic Change*. Belknap Press, 1982.
13 Dowell, G., and A. Swaminathan. 'Racing and Back-Pedalling into the Future: New Product Introduction and Organizational Mortality in the US Bicycle Industry, 1880-1918'. *Organization Studies* 21,

no. 2 (3 January 2000): 405–431.
14 Johnson, G. 'Strategy through a Cultural Lens'. *Management Learning* 31, no. 4 (2000): 403–426.
15 Smit, M. 'London Cars Move No Faster than Chickens', *This is Local London*. Accessed 5 April 2012. http://www.thisislocallondon.co.uk/news/topstories/804876.london_cars_move_no_faster_than_chickens/.
16 Klöckner, C.A., and E. Matthies. 'Two Pieces of the Same Puzzle? Script-Based Car Choice Habits Between the Influence of Socialization and Past Behavior'. *Journal of Applied Social Psychology* 42, no. 4 (2012): 793–821.
17 Gärling, T., S. Fujii, and O. Boe. 'Empirical Tests of a Model of Determinants of Script-based Driving Choice.' *Transportation Research Part F: Traffic Psychology and Behaviour* 4, no. 2 (2001): 89–102.
18 Gärling, T., and K.W. Axhausen. 'Introduction: Habitual Travel Choice'. *Transportation* 30, no. 1 (2003): 1–11.
19 Fujii, S., and R. Kitamura. 'What Does a One-month Free Bus Ticket Do to Habitual Drivers? An Experimental Analysis of Habit and Attitude Change.' *Transportation* 30, no. 1 (2003): 81–95.
20 Eriksson, L., J. Garvill, and A.M. Nordlund. 'Interrupting Habitual Car Use: The Importance of Car Habit Strength and Moral Motivation for Personal Car Use Reduction.' *Transportation Research Part F: Traffic Psychology and Behaviour* 11, no. 1 (2008): 10–23.
21 Wansink, B., and J. Sobal. 'Mindless Eating'. *Environment and Behavior* 39, no. 1 (2007): 106.
22 Wansink, B. *Mindless Eating: Why We Eat More Than We Think*. New York: Bantam, 2007.
23 Wansink, B., and S.B. Park. 'At the Movies: How External Cues and Perceived Taste Impact Consumption Volume'. *Food Quality and Preference* 12, no. 1 (2001): 69–74.
24 Wansink, B., and J. Kim. 'Bad Popcorn in Big Buckets: Portion Size Can Influence Intake as Much as Taste'. *Journal of Nutrition Education and Behavior* 37, no. 5 (2005): 242–245.
25 Mitchell, A. 'Nine American Lifestyles: Values and Societal Change.' *Futurist* 18 (1984): n4.
26 Humby, C., T. Hunt, and T. Phillips. *Scoring Points: How Tesco Continues to Win Customer Loyalty*. London: Kogan Page Ltd, 2008.
27 Ji, M.F., and W. Wood. 'Purchase and Consumption Habits: Not Necessarily What You Intend'. *Journal of Consumer Psychology* 17, no. 4 (2007): 261.
28 Reichheld, F. F., and T. Teal. *The Loyalty Effect: The Hidden Force Behind Growth, Profits, and Lasting Value*. Cambridge, MA: Harvard Business Press, 2001.

29 Chandrashekaran, M., K. Rotte, S.S. Tax, and R. Grewal. 'Satisfaction Strength and Customer Loyalty'. *Journal of Marketing Research* 44, no. 1 (2007): 153–163.
30 Szymanski, D.M., and D.H. Henard. 'Customer Satisfaction: a Meta-analysis of the Empirical Evidence'. *Journal of the Academy of Marketing Science* 29, no. 1 (2001): 16–35.
31 Dijksterhuis, A., P.K. Smith, R.B. Van Baaren, and D.H.J. Wigboldus. 'The Unconscious Consumer: Effects of Environment on Consumer Behavior'. *Journal of Consumer Psychology* 15, no. 3 (2005): 193–202.
32 Murray, K.B., and G. Häubl. 'Explaining Cognitive Lock-In: The Role of Skill-Based Habits of Use in Consumer Choice'. *Journal of Consumer Research* 34, no. 1 (2007): 77–88.
33 Verplanken, B., H. Aarts, and A.D. Van Knippenberg. 'Habit, Information Acquisition, and the Process of Making Travel Mode Choices'. *European Journal of Social Psychology* 27, no. 5 (1997): 539–560.
34 Lal, R., and D.E. Bell. 'The Impact of Frequent Shopper Programs in Grocery Retailing'. *Quantitative Marketing and Economics* 1, no. 2 (2003): 179–202.
35 Wood, W., and D.T. Neal. 'The Habitual Consumer'. *Journal of Consumer Psychology* 19, no. 4 (2009): 579–592.
36 Andreasen, A.R. 'Life Status Changes and Changes in Consumer Preferences and Satisfaction'. *Journal of Consumer Research* 11, no. 3 (1 December 1984): 784–794.
37 Mathur, A., G.P. Moschis, and E. Lee. 'A Longitudinal Study of the Effects of Life Status Changes on Changes in Consumer Preferences'. *Journal of the Academy of Marketing Science* 36, no. 2 (2008): 234–246.

6. Stuck in a depressing loop

1 Rapoport, J.L. *The Boy Who Couldn't Stop Washing: The Experience & Treatment of Obsessive-compulsive Disorder*. New York: Signet, 1992.
2 Karno, M., J.M. Golding, S.B. Sorenson, and M.A. Burnam. 'The Epidemiology of Obsessive-compulsive Disorder in Five US Communities'. *Archives of General Psychiatry* 45, no. 12 (1988): 1094.
3 Gibbs, N.A. 'Nonclinical Populations in Research on Obsessive-compulsive Disorder: A Critical Review.' *Clinical Psychology Review* 16, no. 8 (1996): 729–773.
4 Mancebo, M.C., J.L. Eisen, A. Pinto, B.D. Greenberg, I.R. Dyck, and S.A. Rasmussen. 'The Brown Longitudinal Obsessive Compulsive Study: Treatments Received and Patient Impressions of Improvement'. *Journal of Clinical Psychiatry* 67, no. 11 (2006):

1713–1720.

5 Bentall, R.P. *Madness Explained: Psychosis and Human Nature*. London: Penguin, 2003.

6 Dean, J., '30 Psychobabble Phrases – Which Do You Hate Most?' PsyBlog. http://www.spring.org.uk/2008/06/30-psychobabble-phrases-which-do-you.php

7 Pauls, D.L., K.E. Towbin, J.F. Leckman, G.E.P. Zahner, and D.J. Cohen. 'Gilles De La Tourette's Syndrome and Obsessive-compulsive Disorder: Evidence Supporting a Genetic Relationship.' *Archives of General Psychiatry* 43, no. 12 (1986): 1180.

8 Piacentini, J., D.W. Woods, L. Scahill, S. Wilhelm, A.L. Peterson, S. Chang, G.S. Ginsburg, T. Deckersbach, J. Dziura, and S. Levi-Pearl. 'Behavior Therapy for Children with Tourette Disorder.' *JAMA: Journal of the American Medical Association* 303, no. 19 (2010): 1929.

9 Alloy, L.B., L.Y. Abramson, M.E. Hogan, W.G. Whitehouse, D.T. Rose, M.S. Robinson, R.S. Kim, and J.B. Lapkin. 'The Temple-Wisconsin Cognitive Vulnerability to Depression Project: Lifetime History of Axis I Psychopathology in Individuals at High and Low Cognitive Risk for Depression.' *Journal of Abnormal Psychology* 109, no. 3 (2000): 403.

10 Mezulis, A.H., L.Y. Abramson, J.S. Hyde, and B.L. Hankin. 'Is There a Universal Positivity Bias in Attributions? A Meta-analytic Review of Individual, Developmental, and Cultural Differences in the Self-serving Attributional Bias.' *Psychological Bulletin* 130, no. 5 (2004): 711.

11 Morrow, J., and S. Nolen-Hoeksema. 'Effects of Responses to Depression on the Remediation of Depressive Affect.' *Journal of Personality and Social Psychology* 58, no. 3 (1990): 519.

12 Ward, A., S. Lyubomirsky, L. Sousa, and S. Nolen-Hoeksema. 'Can't Quite Commit: Rumination and Uncertainty.' *Personality and Social Psychology Bulletin* 29, no. 1 (2003): 96–107.

13 Aldao, A., S. Nolen-Hoeksema, and S. Schweizer. 'Emotion-regulation Strategies Across Psychopathology: A Meta-analytic Review.' *Clinical Psychology Review* 30, no. 2 (2010): 217–237.

14 Perkins, A.M., and P.J. Corr. 'Can Worriers Be Winners? The Association Between Worrying and Job Performance.' *Personality and Individual Differences* 38, no. 1 (2005): 25–31.

15 Davey, G.C.L., J. Hampton, J. Farrell, and S. Davidson. 'Some Characteristics of Worrying: Evidence for Worrying and Anxiety as Separate Constructs.' *Personality and Individual Differences* 13, no. 2 (1992): 133–147.

16 Siddique, H.I., V.H. LaSalle-Ricci, C.R. Glass, D.B. Arnkoff, and R.J. Díaz. 'Worry, Optimism, and Expectations as Predictors of

Anxiety and Performance in the First Year of Law School.' *Cognitive Therapy and Research* 30, no. 5 (2006): 667–676.

17 Dijkstra, A., and J. Brosschot. 'Worry About Health in Smoking Behavior Change.' *Behavior Research and Therapy* 41, no. 9 (2003): 1081–1092.

18 Watkins, E.R. 'Constructive and Unconstructive Repetitive Thought.' *Psychological Bulletin* 134, no. 2 (2008): 163.

19 Norem, J.K., and N. Cantor. 'Defensive Pessimism: Harnessing Anxiety as Motivation.' *Journal of Personality and Social Psychology* 51, no. 6 (1986): 1208.

20 For a guide to self-help books on depression, see my article: '6 Self-Help Books for Depression Recommended by Experts', http://www.psyblog.co.uk/2008/01/6-self-help-books-for-depression.php.

7. When bad habits kill

1 US National Transportation Safety Board. 'Delta Air Lines, Boeing 727-232, N473DA. Dallas-Fort Worth International Airport, Texas, 31 August 1988 (Aircraft accident report, NTSB/AAR-89/04).' Washington, DC.: NTSB, 1989.

2 Degani, A., and E.L. Wiener. 'Human Factors of Flight-deck Checklists: The Normal Checklist.' *Design*, May (1991).

3 Boorman, D. 'Today's Electronic Checklists Reduce Likelihood of Crew Errors and Help Prevent Mishaps.' *ICAO Journal* 56, no. 1 (2001): 17–20.

4 Betsch, T., S. Haberstroh, B. Molter, and A. Glockner. 'Oops, I Did It Again – relapse Errors in Routinized Decision Making.' *Organizational Behavior and Human Decision Processes* 93, no. 1 (2004): 62–74.

5 Norman, D.A. 'Categorization of Action Slips.' *Psychological Review* 88, no. 1 (1981): 1.

6 James, W. *Habit*. New York: Henry Holt & Co., 1890.

7 Sloboda, J.A. 'The Effect of Item Position on the Likelihood of Identification by Inference in Prose Reading and Music Reading.' *Canadian Journal of Psychology/Revue Canadienne De Psychologie* 30, no. 4 (1976): 228.

8 Reason, J. 'Actions Not as Planned: The Price of Automatization'. *Aspects of Consciousness* 1 (1979): 67–89.

9 Reason, J.T. *The Human Contribution: Unsafe Acts, Accidents and Heroic Recoveries*. Burlington, VT: Ashgate Publishing, 2008.

10 Streff, F.M., and E.S. Geller. 'Strategies for Motivating Safety Belt Use: The Application of Applied Behavior Analysis'. *Health Education Research* 1, no. 1 (1986): 47–59.

11 Nilsen, P., M. Bourne, and B. Verplanken. 'Accounting for the Role of Habit in Behavioural Strategies for Injury Prevention.' *Inter-*

national Journal of Injury Control & Safety Promotion 15, no. 1 (2008): 33–40.

12 Haddon Jr, W. 'Advances in the Epidemiology of Injuries as a Basis for Public Policy.' *Public Health Reports* 95, no. 5 (1980): 411.

13 Hanson, D., P. Vardon,, and J. Lloyd. 'Safe Communities: An Ecological Approach to Safety Promotion.' (2004). Retrieved 4 December 2011, from http://eprints.jcu.edu.au/1751/4/04Chapters_5-7.pdf

14 Fidler, J.A. Lion Shahab, O. West, M.J. Jarvis, A. McEwen, J.A. Stapleton, E. Vangeli, and R. West. '"The Smoking Toolkit Study": A National Study of Smoking and Smoking Cessation in England.' *BMC Public Health* 11, no. 1 (2011): 479.

15 Gawande, A. *The Checklist Manifesto: How to Get Things Right*. Profile Books, 2010.

16 Aspden, P., J.A. Wolcott, J.L. Bootman and L.R. Cronenwett. *Preventing Medication Errors*. Washington, DC: National Academies Press, 2007.

17 Wolff, A.M., S.A. Taylor, and J.F. McCabe. 'Using Checklists and Reminders in Clinical Pathways to Improve Hospital Inpatient Care.' *Medical Journal of Australia* 181 (2004): 428–431.

18 Auerbach, A.D., H.J. Murff, S.D. Islam. Chapter 23. 'Pre-anesthesia Checklists to Improve Patient Safety. In Markowitz, A.J., K.G. Shojania, B.W. Duncan, K.M. McDonald, and R.M. Wachter (eds). *Making Health Care Safer: A Critical Analysis of Patient Safety Practices*. Evidence Report/Technology Assessment: Number 43. Rockville, MD: Agency for Healthcare Research and Quality, 2001.

19 Hales, B.M., and P.J. Pronovost. 'The Checklist – A Tool for Error Management and Performance Improvement.' *Journal of Critical Care* 21, no. 3 (2006): 231–235.

8. Online all the time

1 Karaiskos, D., E. Tzavellas, G. Balta, and T. Paparrigopoulos. 'P02-232-Social Network Addiction: A New Clinical Disorder?' *European Psychiatry* 25 (2010): 855-855.

2 Jackson, T., R. Dawson, and D. Wilson. *Case Study: Evaluating the Effect of Email Interruptions within the Workplace*. (2002).

3 AOL. 'The 2010 AOL Email Survey.' Accessed 6 January 2012. http://o.aolcdn.com/cdn.webmail.aol.com/survey/aol/en-us/index.htm

4 Czerwinski, M., E. Horvitz, and S. Wilhite. 'A Diary Study of Task Switching and Interruptions.' In *Proceedings of the SIGCHI Conference on Human Factors in Computing Systems*, 175–182. ACM, 2004.

5 Ramsay, J., and K. Renaud. 'Using Insights from Email Users to Inform Organisational Email Management Policy.' *Behaviour &*

Information Technology, no. 1 (2010): 1-1.
6 Einstein, G.O., M.A. McDaniel, C.L. Williford, J.L. Pagan, and R. Dismukes. 'Forgetting of Intentions in Demanding Situations Is Rapid.' *Journal of Experimental Psychology: Applied* 9, no. 3 (2003): 147.
7 Jackson, T., R. Dawson, and D. Wilson. *Case Study: Evaluating the Effect of Email Interruptions Within the Workplace*. Keele, UK: EASE (Empirical Assessment in Software Engineering), 2002.
8 González, V.M., and G. Mark. 'Constant, Constant, Multi-tasking Craziness: Managing Multiple Working Spheres.' *Proceedings of the SIGCHI Conference on Human Factors in Computing Systems*, Vienna, Austria, 24-29 April 2004. New York: ACM (Association for Computing Machinery), 2004, 113–120.
9 Naaman, M., J. Boase, and C.H. Lai. 'Is It Really about Me?: Message Content in Social Awareness Streams.' *Proceedings of the 2010 ACM Conference on Computer Supported Cooperative Work*, Savannah, GA, 6-10 February 2010. New York: ACM, 2010, 189–192.
10 Heil, B., and M. Piskorski. 'New Twitter Research: Men Follow Men and Nobody Tweets.' *Harvard Business Review*, 1 June 2009, http://blogs.hbr.org/cs/2009/06/new_twitter_research_men_follo.html.
11 Johnson, P.R., and S.U. Yang. 'Uses and Gratifications of Twitter: An Examination of User Motives and Satisfaction of Twitter Use.' *Association for Education in Journalism and Mass Communication* (2009).
12 Barnes, S.J., and M. Böhringer. 'Continuance Usage Intention in Microblogging Services: The Case of Twitter.' *Proceedings of the 17th European Conference on Information Systems ECIS*, Verona, Italy, 8-10 June 2009. Padua, Italy: ECIS, 2009, 2:1–13.
13 LaRose, R. 'The Problem of Media Habits.' *Communication Theory* 20, no. 2 (2010): 194–222.
14 LaRose, R., and M.S. Eastin. 'A Social Cognitive Theory of Internet Uses and Gratifications: Toward a New Model of Media Attendance.' *Journal of Broadcasting & Electronic Media* 48, no. 3 (2004): 358–377.
15 LaRose, R., C.A. Lin, and M.S. Eastin. 'Unregulated Internet Usage: Addiction, Habit, or Deficient Self-regulation?' *Media Psychology* 5, no. 3 (2003): 225–253
16 Young, K.S. 'Internet Addiction: The Emergence of a New Clinical Disorder.' *CyberPsychology & Behavior* 1, no. 3 (1998): 237–244.
17 Bergmark, K.H., A. Bergmark, and O. Findahl. 'Extensive Internet Involvement – Addiction or Emerging Lifestyle?' *International Journal of Environmental Research and Public Health* 8, no. 12 (2011): 4488–4501.

18 Salary.com. 'The 2008 Wasting Time at Work Survey Reveals a Record Number of People Waste Time at Work.'. Accessed 4 March 2012, http://www.salary.com/personal/layoutscripts/psnl_articles.asp?tab=psn&cat=cat011&ser=ser033&part=par1083
19 Rideout, V. J., U.G. Foehr, and D.F. Roberts. *Generation M2: Media in the Lives of 8-to 18-year Olds. 2010*. Menlo Park, CA: Henry J Kaiser Family Foundation, 2010.
20 Ophir, E., C. Nass, and A.D. Wagner. 'Cognitive Control in Media Multitaskers.' *Proceedings of the National Academy of Sciences* 106, no. 37 (2009): 15583–15587.

9. Making habits

1 Isaac, B.. 'Jerry Seinfeld's Productivity Secret', *Lifehacker*, 24 July 2007, http://lifehacker.com/281626/jerry-seinfelds-productivity-secret?tag=softwaremotivation
2 Freeman, R. B. *Charles Darwin: a Companion*. Folkestone: Dawson & Sons Ltd., 1978.
3 McCrum, Robert. *Wodehouse: A Life*. New York: W. W. Norton & Company, 2005.
4 Oettingen, G., and D. Mayer. 'The Motivating Function of Thinking about the Future: Expectations versus Fantasies.' *Journal of Personality and Social Psychology* 83, no. 5 (2002): 1198.
5 Pham, L.B., and S.E. Taylor. 'From Thought to Action: Effects of Process-versus outome-based Mental Simulations on Performance.' *Personality and Social Psychology Bulletin* 25, no. 2 (1999): 250–260.
6 Oettingen, G., H. Pak, and K. Schnetter. 'Self-regulation of Goal Setting: Turning Free Fantasies about the Future into Binding Goals.' *Journal of Personality and Social Psychology* 80, no. 5 (May 2001): 736–753.
7 Oettingen, G. 'Future Thought and Behaviour Change.' *European Review of Social Psychology* 23, no. 1 (2012): 1–63.
8 Chapman, J., C.J. Armitage, and P. Norman. 'Comparing Implementation Intention Interventions in Relation to Young Adults' Intake of Fruit and Vegetables.' *Psychology and Health* 24, no. 3 (2009): 317–332.
9 Gollwitzer, P.M., and P. Sheeran. 'Implementation Intentions and Goal Achievement: A Meta-analysis of Effects and Processes.' *Advances in Experimental Social Psychology* 38 (2006): 69–119.
10 Gollwitzer, P.M., F. Wieber, A.L. Meyers, and S.M. McCrea. 'How to Maximize Implementation Intention Effects.' In Agnew, C.R., D.E. Carlston, W.G. Graziano, and J.R. Kelly (eds). *Then a Miracle Occurs: Focusing on Behaviour in Social Psychological Theory and Research* (2010): 137–161.
11 McDaniel, M.A., and G.O. Einstein. 'Strategic and Automatic

Processes in Prospective Memory Retrieval: A Multiprocess Framework.' *Applied Cognitive Psychology* 14, no. 7 (2000): S127-S144.
12 Lally, P., J. Wardle, and B. Gardner. 'Experiences of Habit Formation: A Qualitative Study.' *Psychology, Health & Medicine* 16, no. 4 (2011): 484–489.
13 Graybiel, A.M. 'The Basal Ganglia and Chunking of Action Repertoires.' *Neurobiology of Learning and Memory* 70, no. 1–2 (July 1998): 119–136.
14 Gollwitzer, P.M, F. Wieber, A.L. Myers, and S.M. McCrea. 'How to Maximize Implementation Intention Effects.' *Then A Miracle Occurs,* eds. C. R. Agnew, D. E. Carlston, W. G. Graziano, and J. R. Kelly. New York: Oxford University Press, 2009.
15 Webb, T.L., J. Christian, and C.J. Armitage. 'Helping Students Turn up for Class: Does Personality Moderate the Effectiveness of an Implementation Intention Intervention?' *Learning and Individual Differences* 17, no. 4 (2007): 316–327.
16 Achtziger, A., P.M. Gollwitzer, and P. Sheeran. 'Implementation Intentions and Shielding Goal Striving from Unwanted Thoughts and Feelings.' *Personality and Social Psychology Bulletin* 34, no. 3 (2008): 381–393.
17 Osch, L., L. Lechner, A. Reubsaet, S. Wigger, and H. Vries. 'Relapse Prevention in a National Smoking Cessation Contest: Effects of Coping Planning.' *British Journal of Health Psychology* 13, no. 3 (2008): 525–535.
18 Scholz, U., B. Schüz, J.P. Ziegelmann, S. Lippke, and R. Schwarzer. 'Beyond Behavioural Intentions: Planning Mediates between Intentions and Physical Activity.' *British Journal of Health Psychology* 13, no. 3 (2008): 479–494.
19 Burke, L.E., V. Swigart, M. Warziski Turk, N. Derro, and L.J. Ewing. 'Experiences of Self-Monitoring: Successes and Struggles during Treatment for Weight Loss.' *Qualitative Health Research* 19, no. 6 (June 2009): 815–828.
20 Deci, E.L., and R.M. Ryan. *Intrinsic Motivation and Self-determination in Human Behaviour (Perspectives in Social Psychology).* Berlin: Springer, 1985.

10. Breaking habits

1 Norcross, J.C., A.C. Ratzin, and D. Payne. 'Ringing in the New Year: The Change Processes and Reported Outcomes of Resolutions.' *Addictive Behaviors* 14, no. 2 (1989): 205–212.
2 Shapiro, S.L., L.E. Carlson, J.A. Astin, and B. Freedman. 'Mechanisms of Mindfulness'. *Journal of Clinical Psychology* 62, no. 3 (2006): 373–386.
3 Chatzisarantis, N.L.D., and M.S. Hagger. 'Mindfulness and the

Intention-behaviour Relationship Within the Theory of Planned Behavior'. *Personality and Social Psychology Bulletin* 33, no. 5 (2007): 663–676.

4 Quinn, J.M., A. Pascoe, W. Wood, and D.T. Neal. 'Can't Control Yourself? Monitor Those Bad Habits.' *Personality and Social Psychology Bulletin* 36, no. 4 (2010): 499–511.

5 Wegner, D.M., D.J. Schneider, S.R. Carter, and T.L. White. 'Paradoxical Effects of Thought Suppression.' *Journal of Personality and Social Psychology* 53, no. 1 (1987): 5.

6 Salkovskis, P.M., and M. Reynolds. 'Thought Suppression and Smoking Cessation.' *Behaviour Research and Therapy* 32, no. 2 (February 1994): 193–201.

7 Polivy, J. 'The Effects of Behavioural Inhibition: Integrating Internal Cues, Cognition, Behaviour, and Affect.' *Psychological Inquiry* 9, no. 3 (1998): 181–204.

8 Polivy, J., and C.P. Herman. 'Dieting and Binging: A Causal Analysis.' *American Psychologist* 40, no. 2 (1985): 193.

9 Johnson, F., M. Pratt, and J. Wardle. 'Dietary Restraint and Self-regulation in Eating Behavior.' *International Journal of Obesity* 36, no. 5 (2012): 665–674.

10 Marlatt, G.A, and D.M. Donovan. *Relapse Prevention: Maintenance Strategies in the Treatment of Addictive Behaviors*. New York: Guilford Press, 2005.

11 Collins, R.L, and W.M Lapp. 'The Temptation and Restraint Inventory for measuring drinking restraint.' *British Journal of Addiction* 87, no. 4 (1992): 625–633.

12 Wood, W., and D.T. Neal. 'A New Look at Habits and the Habit-goal Interface.' *Psychological Review* 114, no. 4 (2007): 843.

13 Bouton, M.E. 'Context, Ambiguity, and Unlearning: Sources of Relapse after Behavioural Extinction.' *Biological Psychiatry* 52, no. 10 (2002): 976–986.

14 Betsch, T., S. Haberstroh, B. Molter, and A. Glockner. 'Oops, I Did It Again – Relapse Errors in Routinized Decision Making.' *Organizational Behavior and Human Decision Processes* 93, no. 1 (2004): 62-74.

15 Cohen, A.L., U.C. Bayer, A. Jaudas, and P.M. Gollwitzer. 'Self-regulatory Strategy and Executive Control: Implementation Intentions Modulate Task Switching and Simon Task Performance.' *Psychological Research* 72, no. 1 (2008): 12–26.

16 Stewart, B.D., and B.K. Payne. 'Bringing Automatic Stereotyping Under Control: Implementation Intentions as Efficient Means of Thought Control.' *Personality and Social Psychology Bulletin* 34, no. 10 (2008): 1332–1345.

17 Nordgren, L.F., R. van Harreveld, and J. van der Pligt. 'The Restraint Bias: How the Illusion of Self-restraint Promotes Impul-

sive Behavior.' *Psychological Science* 20, no. 12 (December 2009): 1523–1528.

18 Baumeister, R.F., E. Bratslavsky, M. Muraven, and D.M. Tice. 'Ego Depletion: Is the Active Self a Limited Resource?' *Journal of Personality and Social Psychology* 74, no. 5 (1998): 1252.

19 Muraven, M., and R.F. Baumeister. 'Self-regulation and Depletion of Limited Resources: Does Self-control Resemble a Muscle?' *Psychological Bulletin* 126, no. 2 (2000): 247.

20 Ariely, D., and K. Wertenbroch. 'Procrastination, Deadlines, and Performance: Self-control by Precommitment.' *Psychological Science* 13, no. 3 (May 2002): 219–224.

21 Thaler, R.H., and S. Benartzi. 'Save More Tomorrow™: Using Behavioral Economics to Increase Employee Saving.' *Journal of Political Economy* 112, no. S1 (2004): S164–S187.

22 Trope, Y., and A. Fishbach. 'Counteractive Self-control in Overcoming Temptation.' *Journal of Personality and Social Psychology* 79, no. 4 (2000): 493–506.

23 Zhang, Y., and A. Fishbach. 'Counteracting Obstacles with Optimistic Predictions.' *Journal of Experimental Psychology* 139, no. 1 (February 2010): 16–31.

24 Fishbach, A., Y. Zhang, and Y. Trope. 'Counteractive Evaluation: Asymmetric Shifts in the Implicit Value of Conflicting Motivations.' *Journal of Experimental Social Psychology* 46, no. 1 (January 2010): 29–38.

25 Schmeichel, B.J., and K. Vohs. 'Self-affirmation and Self-control: Affirming Core Values Counteracts Ego Depletion.' *Journal of Personality and Social Psychology* 96, no. 4 (2009): 770.

26 Fujita, K., and J.C. Roberts. 'Promoting Prospective Self-control through Abstraction.' *Journal of Experimental Social Psychology* 46, no. 6 (2010): 1049–1054.

27 Oaten, M., and K. Cheng. 'Longitudinal Gains in Self-regulation from Regular Physical Exercise.' *British Journal of Health Psychology* 11, no. 4 (2006): 717–733.

28 Oaten, M., and K. Cheng. 'Improvements in Self-control from Financial Monitoring.' *Journal of Economic Psychology* 28, no. 4 (2007): 487–501.

29 Wood, W., L. Tam, and M.G. Witt. 'Changing circumstances, disrupting habits.' *Journal of Personality and Social Psychology* 88, no. 6 (2005): 918.

30 Verplanken, B., I. Walker, A. Davis, and M. Jurasek. 'Context Change and Travel Mode Choice: Combining the Habit Discontinuity and Self-activation Hypotheses.' *Journal of Environmental Psychology* 28, no. 2 (2008): 121–127.

31 Soler, R.E., K.D. Leeks, L.R. Buchanan, R.C. Brownson, G.W.

Heath, and D.H. Hopkins. 'Point-of-Decision Prompts to Increase Stair Use: A Systematic Review Update.' *American Journal of Preventive Medicine* 38, no. 2 (2010): S292–S300.

32 Tobias, R. 'Changing Behavior by Memory Aids: A Social Psychological Model of Prospective Memory and Habit Development Tested with Dynamic Field Data.' *Psychological Review* 116, no. 2 (2009): 408.

33 Münsterberg, H. 'Gedächtnisstudien.' *Beiträge Zur Experimentellen Psychologie* 4 (1892): 70.

11. Healthy habits

1 Ogden, C.L., and M.D. Carroll. *Prevalence of Overweight, Obesity, and Extreme Obesity Among Adults: United States, Trends 1960-1962 Through 2007-2008*. NCHS Health E-Stats. http://www.cdc.gov/nchs/data/hestat/obesity_adult_07_08/obesity_adult_07_08.htm. Accessed 12 April 2012.

2 Finucane, M.M., G.A. Stevens, M.J. Cowan, G. Danaei, J.K. Lin, C.J. Paciorek, and G.M. Singh. 'National, Regional, and Global Trends in Body-mass Index Since 1980: Systematic Analysis of Health Examination Surveys and Epidemiological Studies with 960 Country-years and 9·1 Million Participants.' *The Lancet* 377, no. 9765 (12): 557–567.

3 Wing, R.R., and S. Phelan. 'Long-Term Weight Loss Maintenance.' *American Journal of Clinical Nutrition* 82, no. 1 (1 July 2005): 222S–225S.

4 Snyder, L.B., M.A. Hamilton, E.W. Mitchell, J. Kiwanuka-Tondo, F. Fleming-Milici, and D. Proctor. 'A Meta-analysis of the Effect of Mediated Health Communication Campaigns on Behavior Change in the United States'. *Journal of Health Communication* 9, no. S1 (2004): 71–96.

5 Riet, J., S.J. Sijtsema, H. Dagevos, and G.J. De Bruijn. 'The Importance of Habits in Eating Behaviour: An Overview and Recommendations for Future Research'. *Appetite* 57, no. 3 (2011): 585–596.

6 Gardner, B., G.J. de Bruijn, and P. Lally. 'A Systematic Review and Meta-analysis of Applications of the Self-Report Habit Index to Nutrition and Physical Activity Behaviours'. *Annals of Behavioral Medicine* (2011): 1–14.

7 Ji, M.F., and W. Wood. 'Purchase and Consumption Habits: Not Necessarily What You Intend'. *Journal of Consumer Psychology* 17, no. 4 (2007): 261.

8 Neal, D.T, W. Wood, M. Wu, and D. Kurlander. 'The Pull of the Past'. *Personality and Social Psychology Bulletin* 37, no. 11 (2011): 1428–1437.

9 Daeninck, E., and M. Miller. 'What Can the National Weight Control

Registry Teach Us?' *Current Diabetes Reports* 6, no. 5 (2006): 401–404.

10 Wing, R.R., and S. Phelan. 'Long-term Weight Loss Maintenance.' *American Journal of Clinical Nutrition* 82, no. 1 (2005): 222S–225S.

11 Chandon, P., and B. Wansink. 'When Are Stockpiled Products Consumed Faster? A Convenience-salience Framework of Postpurchase Consumption Incidence and Quantity'. *Journal of Marketing Research* (2002): 321–335.

12 Wansink, B., J.E. Painter, and Y.K. Lee. 'The Office Candy Dish: Proximity's Influence on Estimated and Actual Consumption'. *International Journal of Obesity* 30, no. 5 (2006): 871–875.

13 Wansink, B., and M.M. Cheney. 'Super Bowls: Serving Bowl Size and Food Consumptio.n'. *JAMA: Journal of the American Medical Association* 293, no. 14 (2005): 1727.

14 Lawless, H.T., S. Bender, C. Oman, and C. Pelletier. 'Gender, Age, Vessel Size, Cup Vs. Straw Sipping, and Sequence Effects on Sip Volume'. *Dysphagia* 18, no. 3 (2003): 196–202.

15 Sobal, J., and B. Wansink. 'Kitchenscapes, Tablescapes, Platescapes, and Foodscapes'. *Environment and Behavior* 39, no. 1 (2007): 124–142.

16 Papies, E.K., L.W. Barsalou, and R. Custers. 'Mindful Attention Prevents Mindless Impulses.' *Social Psychological and Personality Science* 3, no. 3 (1 May 2012): 291–299.

17 Adriaanse, M.A., J.M.F. van Oosten, D.T.D. de Ridder, J.B.F. de Wit, and C. Evers. 'Planning What Not to Eat: Ironic Effects of Implementation Intentions Negating Unhealthy Habits'. *Personality and Social Psychology Bulletin* 37, no. 1 (2011): 69.

18 Adriaanse, M.A., G. Oettingen, P.M. Gollwitzer, E.P. Hennes, D.T.D. de Ridder, and J.B.F. de Wit. 'When Planning Is Not Enough: Fighting Unhealthy Snacking Habits by Mental Contrasting with Implementation Intentions (MCII)'. *European Journal of Social Psychology* 40, no. 7 (2010): 1277–1293.

19 Ibid., p. 1281

20 Smith, P.J., J.A. Blumenthal, B.M. Hoffman, H. Cooper, T.A. Strauman, K. Welsh-Bohmer, J.N. Browndyke, and A. Sherwood. 'Aerobic Exercise and Neurocognitive Performance: a Meta-Analytic Review of Randomized Controlled Trials.' *Psychosomatic Medicine* 72, no. 3 (April 2010): 239–252.

21 Stathopoulou, G., M.B. Powers, A.C. Berry, J.A.J. Smits, and M.W. Otto. 'Exercise Interventions for Mental Health: A Quantitative and Qualitative Review.' *Clinical Psychology: Science and Practice* 13, no. 2 (1 May 2006): 179–193.

22 Babyak, M., J.A. Blumenthal, S. Herman, P. Khatri, M. Doraiswamy, K. Moore, W.E. Craighead, T.T. Baldewicz, and K.R. Krishnan.

'Exercise Treatment for Major Depression: Maintenance of Therapeutic Benefit at 10 Months.' *Psychosomatic Medicine* 62, no. 5 (2000): 633–638.

23 Gardner, B., G.J. de Bruijn, and P. Lally. 'A Systematic Review and Meta-analysis of Applications of the Self-Report Habit Index to Nutrition and Physical Activity Behaviours'. *Annals of Behavioral Medicine* (2011): 1–14.

24 Yang, X., R. Telama, M. Leino, and J. Viikari. 'Factors Explaining the Physical Activity of Young Adults: The Importance of Early Socialization'. *Scandinavian Journal of Medicine & Science in Sports* 9, no. 2 (1999): 120–127.

25 Michie, S., C. Abraham, C. Whittington, J. McAteer, and S. Gupta. 'Effective Techniques in Healthy Eating and Physical Activity Interventions: A Meta-regression.' *Health Psychology* 28, no. 6 (2009): 690.

26 Baker, G., S.R., Gray, A. Wright, C. Fitzsimons, M. Nimmo, R. Lowry, N. Mutrie, and the Scottish Physical Activity Research Collaboration (SPARColl). 'The Effect of a Pedometer-based Community Walking Intervention'. *International Journal of Behavioral Nutrition and Physical Activity* 5, no. 1 (5 September 2008): 44.

27 Armitage, C.J., and M.A. Arden. 'A Volitional Help Sheet to Increase Physical Activity in People with Low Socioeconomic Status: A Randomised Exploratory Trial.' *Psychology and Health* 25, no. 10 (2010): 1129–1145.

28 West, R., and J. Fidler. *Smoking and Smoking Cessation in England 2010*. London: Vasco-Graphics, 2011.

29 Taylor, T., and UK National Statistics. *Smoking-related Behaviour and Attitudes, 2005*. London: Office for National Statistics, 2006.

30 Balfour, D.J.K. 'The Neurobiology of Tobacco Dependence: A Preclinical Perspective on the Role of the Dopamine Projections to the Nucleus'. *Nicotine & Tobacco Research* 6, no. 6 (1 December 2004): 899–912.

31 Hughes, J.R., L.F. Stead, and T. Lancaster. 'Antidepressants for Smoking Cessation'. *Cochrane Database of Systematic Reviews* 1, no. 1 (24 January 2007): CD000031.

32 Stead, L.F., R. Perera, C. Bullen, D. Mant, and T. Lancaster. 'Nicotine Replacement Therapy for Smoking Cessation.' *Cochrane Database of Systematic Reviews* 1, no. 1 (23 January 2008): CD000146.

33 Orbell, S., and B. Verplanken. 'The Automatic Component of Habit in Health Behavior: Habit as Cue-contingent Automaticity.' *Health Psychology* 29, no. 4 (2010): 374.

34 Michie, S., N. Hyder, A. Walia, and R. West. 'Development of a Taxonomy of Behaviour Change Techniques Used in Individual Behavioural Support for Smoking Cessation'. *Addictive Behaviors*

36, no. 4 (2010): 315–319.
35 Oettingen, G., D. Mayer, and J. Thorpe. 'Self-regulation of Commitment to Reduce Cigarette Consumption: Mental Contrasting of Future with Reality'. *Psychology and Health* 25, no. 8 (2010): 961–977.
36 Mottillo, S., K.B. Filion, P. Bélisle, L. Joseph, A. Gervais, J. O'Loughlin, G. Paradis, R. Pihl, L. Pilote, and S. Rinfret. 'Behavioural Interventions for Smoking Cessation: a Meta-analysis of Randomized Controlled Trials'. *European Heart Journal* 30, no. 6 (2009): 718.
37 Armitage, C.J. 'A Volitional Help Sheet to Encourage Smoking Cessation: A Randomized Exploratory Trial.' *Health Psychology* 27, no. 5 (2008): 557.
38 Conner, M., and A.R. Higgins. 'Long-term Effects of Implementation Intentions on Prevention of Smoking Uptake Among Adolescents: A Cluster Randomized Controlled Trial.' *Health Psychology* 29, no. 5 (2010): 529–538.
39 West, R., A. Walia, N. Hyder, L. Shahab, and S. Michie. 'Behavior Change Techniques Used by the English Stop Smoking Services and Their Associations with Short-term Quit Outcomes'. *Nicotine & Tobacco Research* 12, no. 7 (2010): 742–747.
40 Fidler et al. (2011)
41 Orbell, S., P. Lidierth, C.J. Henderson, N. Geeraert, C. Uller, A.K. Uskul, and M. Kyriakaki. 'Social–cognitive Beliefs, Alcohol, and Tobacco Use: A Prospective Community Study of Change Following a Ban on Smoking in Public Places.' *Health Psychology* 28, no. 6 (2009): 753.
42 Fichtenberg, C.M. 'Effect of Smoke-free Workplaces on Smoking Behaviour: Systematic Review.' *British Medical Journal*, 325, no. 7357 (27 July 2002): 188–188.
43 Anger, S., M. Kvasnicka, and T. Siedler. 'One Last Puff? Public Smoking Bans and Smoking Behavior.' *Journal of Health Economics* 30, no. 3 (May 2011): 591–601.

12. The creative habit

1 Ghiselin, B. *The Creative Process*. Signet, 1952.
2 Maier, N.R.F. 'Reasoning in Humans. II. The Solution of a Problem and Its Appearance in Consciousness.' *Journal of Comparative Psychology* 12, no. 2 (1931): 181.
3 Dror, I.E. 'The Paradox of Human Expertise: Why Experts Can Get It Wrong.' In *The Paradoxical Brain*, ed. Narinder Kapur. Cambridge, UK: Cambridge University Press, 2011.
4 Wiley, J. 'Expertise as Mental Set: The Effects of Domain Knowledge in Creative Problem Solving.' *Memory & Cognition* 26, no. 4

(1998): 716.

5 Yokochi, S., and T. Okada. 'Creative Cognitive Process of Art Making: A Field Study of a Traditional Chinese Ink Painter.' *Creativity Research Journal* 17, no. 2 (2005): 241–255.

6 Finke, R.A., T.B. Ward, and S.M. Smith. *Creative Cognition: Theory, Research, and Applications*. Cambridge, MA: MIT Press, 1992.

7 Markman, K.D., M.J. Lindberg, L.J. Kray, and A.D. Galinsky. 'Implications of Counterfactual Structure for Creative Generation and Analytical Problem Solving.' *Personality and Social Psychology Bulletin* 33, no. 3 (2007): 312.

8 Getzels, J.W., and M. Csikszentmihalyi. *The Creative Vision: A Longitudinal Study of Problem Finding in Art*. Wiley New York, 1976.

9 Rostan, S.M. 'Problem Finding, Problem Solving, and Cognitive Controls: An Empirical Investigation of Critically Acclaimed Productivity.' *Creativity Research Journal* 7, no. 2 (1994): 97–110.

10 Rothenberg, A. 'The Janusian Process in Scientific Creativity.' *Creativity Research Journal* 9, no. 2 (1996): 207–231.

11 Clement, C.A., R. Mawby, and D.E. Giles. 'The Effects of Manifest Relational Similarity on Analog Retrieval.' *Journal of Memory and Language* 33, no. 3 (1994): 396–420.

12 Loewenstein, J. 'How One's Hook Is Baited Matters for Catching an Analogy.' *Psychology of Learning and Motivation: Advances in Research and Theory* (2010): 149.

13 Gassmann, O., and M. Zeschky. 'Opening up the Solution Space: The Role of Analogical Thinking for Breakthrough Product Innovation.' *Creativity and Innovation Management* 17, no. 2 (2008): 97–106.

14 Förster, J., K. Epstude, and A. Özelsel. 'Why Love Has Wings and Sex Has Not: How Reminders of Love and Sex Influence Creative and Analytic Thinking.' *Personality and Social Psychology Bulletin* 35, no. 11 (2009): 1479.

15 Förster, J., R.S. Friedman, and N. Liberman. 'Temporal Construal Effects on Abstract and Concrete Thinking: Consequences for Insight and Creative Cognition* 1.' *Journal of Personality and Social Psychology* 87, no. 2 (2004): 177–189.

16 Jia, L., E.R. Hirt, and S.C. Karpen. 'Lessons from a Faraway Land: The Effect of Spatial Distance on Creative Cognition.' *Journal of Experimental Social Psychology* 45, no. 5 (2009): 1127–1131.

17 Gratzer, W. *Eurekas and Euphorias: The Oxford Book of Scientific Anecdotes*. New York: Oxford University Press, 2004.

18 Wotiz, J.H. *The Kekulé Riddle: A Challenge*. Carbondale, IL: Glenview Press, 1992.

19 Harris, P.L. *The Work of the Imagination. Understanding Children's Worlds*. Malden, MA: Blackwell, 2000.

20 Zabelina, D.L., and M.D. Robinson. 'Child's Play: Facilitating the Originality of Creative Output by a Priming Manipulation.' *Psychology of Aesthetics, Creativity, and the Arts* 4, no. 1 (2010): 57–65.
21 Zabelina, D.L., and M.D. Robinson. 'Creativity as Flexible Cognitive Control.' *Psychology of Aesthetics, Creativity, and the Arts* 4, no. 3 (2010): 136–143.
22 Vartanian, O. 'Variable Attention Facilitates Creative Problem Solving.' *Psychology of Aesthetics, Creativity, and the Arts* 3, no. 1 (2009): 57–59.

13. Happy habits

1 *The Simpsons*. Episode no. 505, first broadcast 29 April 2012 by Fox. Directed by Chris Clements and written by Matt Warburton.
2 Sheldon, K.M, and S. Lyubomirsky. 'Is It Possible to Become Happier? (And If so, How?)'. *Social and Personality Psychology Compass* 1, no. 1 (2007): 129–145.
3 Lykken, D., and A. Tellegen. 'Happiness Is a Stochastic Phenomenon.' *Psychological Science* 7, no. 3 (1996): 186.
4 Myers, D.G. 'The Funds, Friends, and Faith of Happy People.' *American Psychologist* 55, no. 1 (2000): 56.
5 Wilson, T.D., D.T. Gilbert, and D.B. Centerbar. 'Making Sense: The Causes of Emotional Evanescence.' In *The Psychology of Economic Decisions*, ed. Isabelle Brocas and Juan D. Carrillo. New York: Oxford University Press, 2003.
6 Brickman, P., D. Coates, and R. Janoff-Bulman. 'Lottery Winners and Accident Victims: Is Happiness Relative?' *Journal of Personality and Social Psychology* 36, no. 8 (1978): 917.
7 Clark, A.E. 'Are Wages Habit-forming? Evidence from Micro Data'. *Journal of Economic Behavior & Organization* 39, no. 2 (1999): 179–200.
8 Easterlin, R., and L. Angelescu. 'Happiness and Growth the World over: Time Series Evidence on the Happiness-income Paradox.' IZA Discussion Paper No. 4060 (2009).
9 Gilbert, D.T., E.C. Pinel, T.D. Wilson, S.J. Blumberg, and T.P. Wheatley. 'Immune Neglect: A Source of Durability Bias in Affective Forecasting.' *Journal of Personality and Social Psychology* 75, no. 3 (1998): 617.
10 Meyvis, T., R.K. Ratner, and J. Levav. 'Why Don't We Learn to Accurately Forecast Feelings? How Misremembering Our Predictions Blinds Us to Past Forecasting Errors.' *Journal of Experimental Psychology: General; Journal of Experimental Psychology: General* 139, no. 4 (2010): 579.
11 Lyubomirsky, S. *The How of Happiness: A New Approach to Getting the Life You Want*. New York: Penguin, 2008.

12 Emmons, R.A., and M.E. McCullough. 'Counting Blessings Versus Burdens: An Experimental Investigation of Gratitude and Subjective Well-being in Daily Life.' *Journal of Personality and Social Psychology* 84, no. 2 (2003): 377.
13 Larsen, R. 'The Contributions of Positive and Negative Affect to Emotional Well-being.' *Psychological Topics* 18, no. 2 (2009): 247–266.
14 Sheldon, K.M., and S. Lyubomirsky. ' Change Your Actions, Not Your Circumstances: An Experimental Test of the Sustainable Happiness Model'. *Happiness, Economics and Politics: Towards a Multidisciplinary Approach* (2009): 324.
15 Frattaroli, J. 'Experimental Disclosure and Its Moderators: A Meta-analysis.' *Psychological Bulletin* 132, no. 6 (2006): 823.
16 King, L.A. 'The Health Benefits of Writing About Life Goals'. *Personality and Social Psychology Bulletin* 27, no. 7 (2001): 798.
17 Sheldon, K.M., and S. Lyubomirsky. 'How to Increase and Sustain Positive Emotion: The Effects of Expressing Gratitude and Visualizing Best Possible Selves'. *Journal of Positive Psychology* 1, no. 2 (2006): 73–82.
18 Quoidbach, J., E.V. Berry, M. Hansenne, and M. Mikolajczak. 'Positive Emotion Regulation and Well-being: Comparing the Impact of Eight Savoring and Dampening Strategies'. *Personality and Individual Differences* 49, no. 5 (October 2010): 368–373.
19 Dunn, E.W., L.B. Aknin, and M.I. Norton. 'Spending Money on Others Promotes Happiness'. *Science* 319, no. 5870 (2008): 1687.
20 Seligman, M.E.P., T.A. Steen, N. Park, and C. Peterson. 'Positive Psychology Progress: Empirical Validation of Interventions.' *American Psychologist* 60, no. 5 (2005): 410.
21 Killingsworth, M.A., and D.T. Gilbert. 'A Wandering Mind Is an Unhappy Mind'. *Science* 330, no. 6006 (2010): 932.

INDEX

Index

ABOUT THE AUTHOR

Psychologist Jeremy Dean is the founder and author of the website PsyBlog (www.psyblog.co.uk), which receives upwards of 1 million hits monthly and has been featured on the BBC and in the *Guardian*, *The Times*, the *Huffington Post*, and many others. A former lawyer and Internet entrepreneur who changed careers out of a fascination for how the human mind works, he is currently a researcher and PhD candidate at University College London. He lives in Kent.